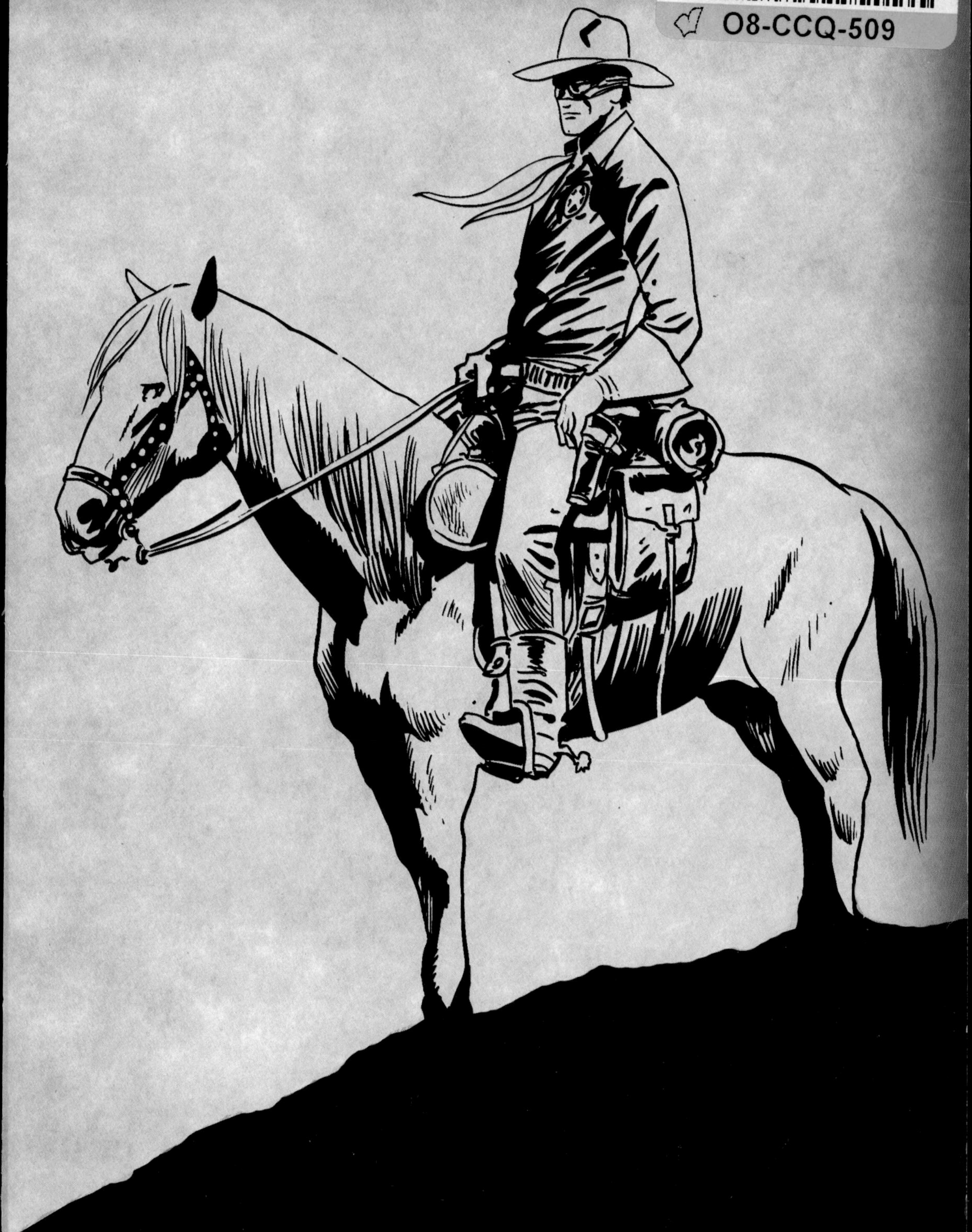

THE LONE RANGER®

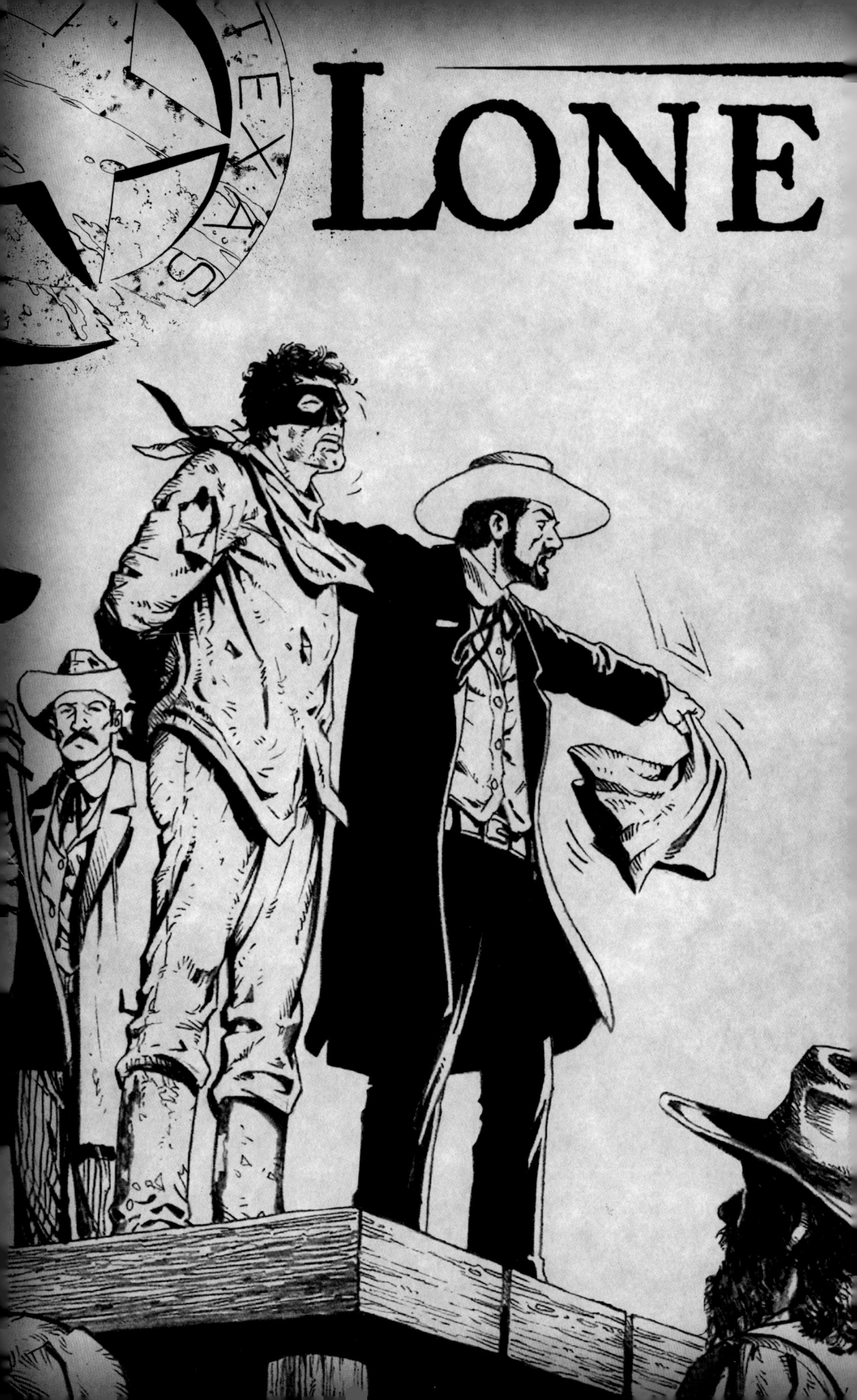
TEXAS
LONE

THE RANGER®

★ VOLUME V: HARD COUNTRY ★

WRITER:
ANDE PARKS

ARTIST:
ESTEVE POLLS

COLORIST:
MARCELO PINTO
OF IMPACTO STUDIO

COVER ARTIST:
ALEX ROSS

COLLECTION DESIGN:
JASON ULLMEYER

SPECIAL THANKS TO:
SCOTT SHILLET
COLIN MCLAUGHLIN
DAMIEN TROMEL

THIS VOLUME COLLECTS THE LONE RANGER VOLUME 2 ISSUES 1-6
BY DYNAMITE ENTERTAINMENT

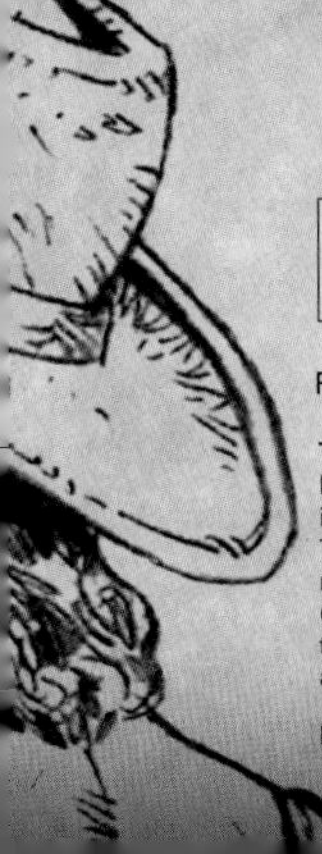

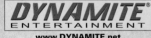

DYNAMITE® ENTERTAINMENT
www.DYNAMITE.net
Follow us on Twitter @dynamitecomics

Nick Barrucci, President
Juan Collado, Chief Operating Officer
Joe Rybandt, Editor
Josh Johnson, Creative Director
Rich Young, Director Business Development
Jason Ullmeyer, Senior Designer
Josh Green, Traffic Coordinator
Chris Caniano, Production Assistant

First Printing
ISBN-10: 1-60690-346-2
ISBN-13: 978-1-60690-346-9
10 9 8 7 6 5 4 3 2 1

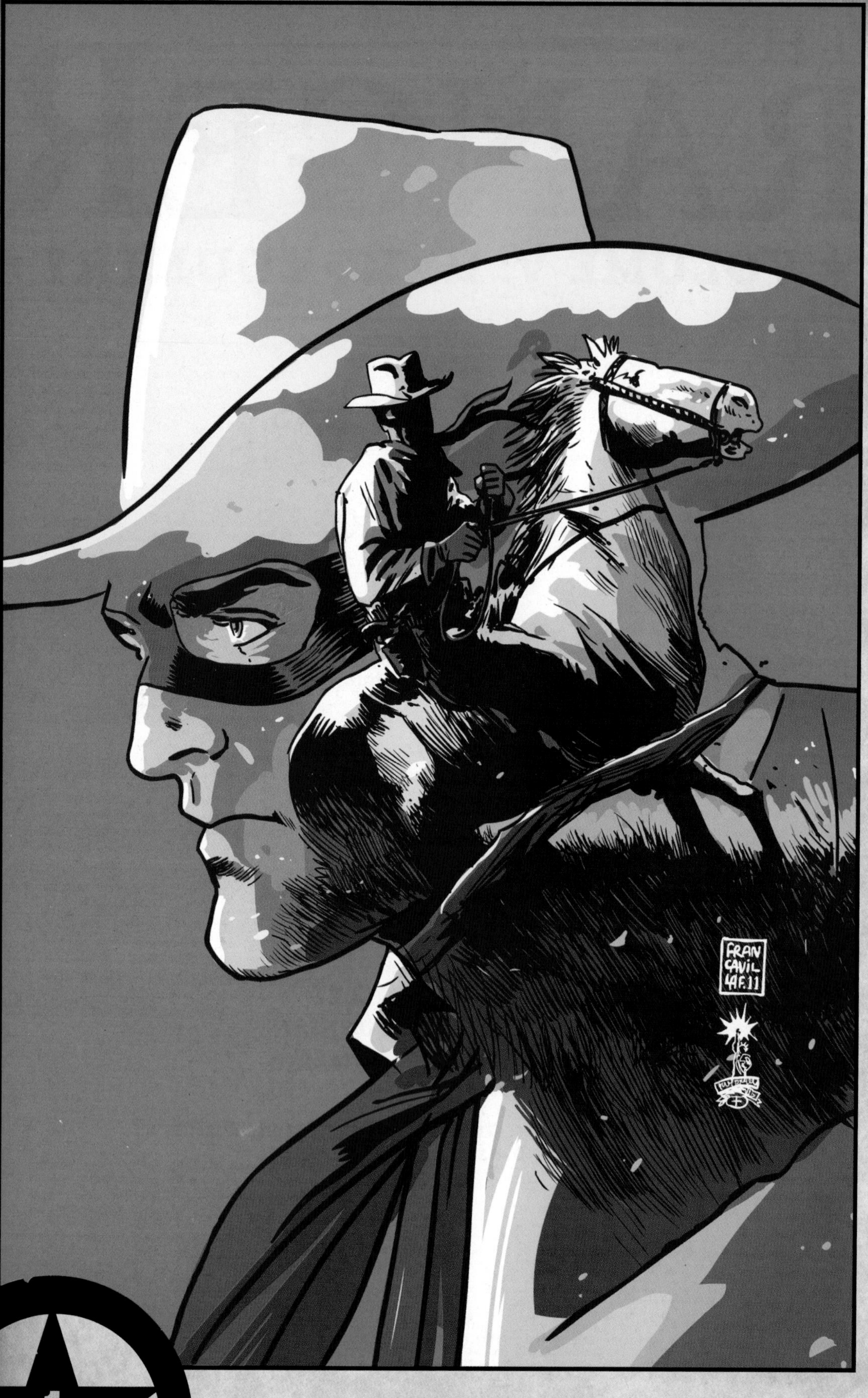
FRAN
CAVIL
LA F. 11

1

COVER BY **FRANCESCO FRANCAVILLA**

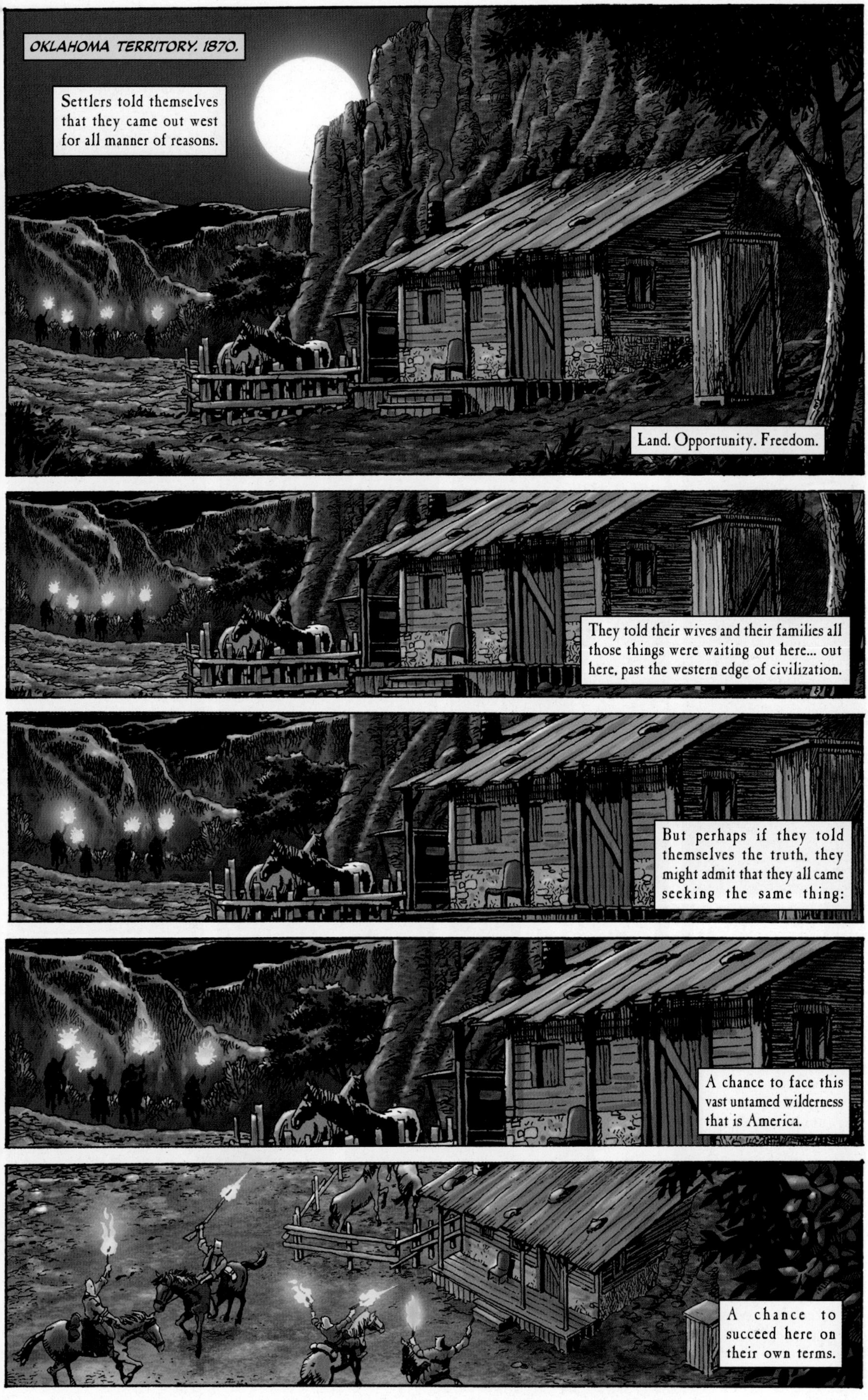
OKLAHOMA TERRITORY. 1870.
Settlers told themselves that they came out west for all manner of reasons.
Land. Opportunity. Freedom.
They told their wives and their families all those things were waiting out here... out here, past the western edge of civilization.
But perhaps if they told themselves the truth, they might admit that they all came seeking the same thing:
A chance to face this vast untamed wilderness that is America.
A chance to succeed here on their own terms.

Few of them thrived as they had dreamed when they first hitched up their wagons and headed their horses west.

WE *KNOW* YOU GOT A LOCK BOX, JURGENS!

GET IT *OUT* HERE!

Many were met with illness, starvation, poverty, and an early grave.

GET IT *NOW* OR WE LIGHT THIS PLACE-- *HUKK!*

A grave dug into these very fields they came here to tame.

NATHAN!

MARTHA! GET *BACK* IN--

The West broke many dreams and many lives.

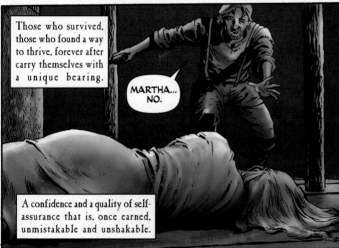

Those who survived, those who found a way to thrive, forever after carry themselves with a unique bearing.

MARTHA... NO.

A confidence and a quality of self-assurance that is, once earned, unmistakable and unshakable.

NO...

It is a quality one never forgets or fails to admire, for it has been, without fail, harshly earned.

Editor, Lincoln County News - Lincoln, Kansas, 1882.

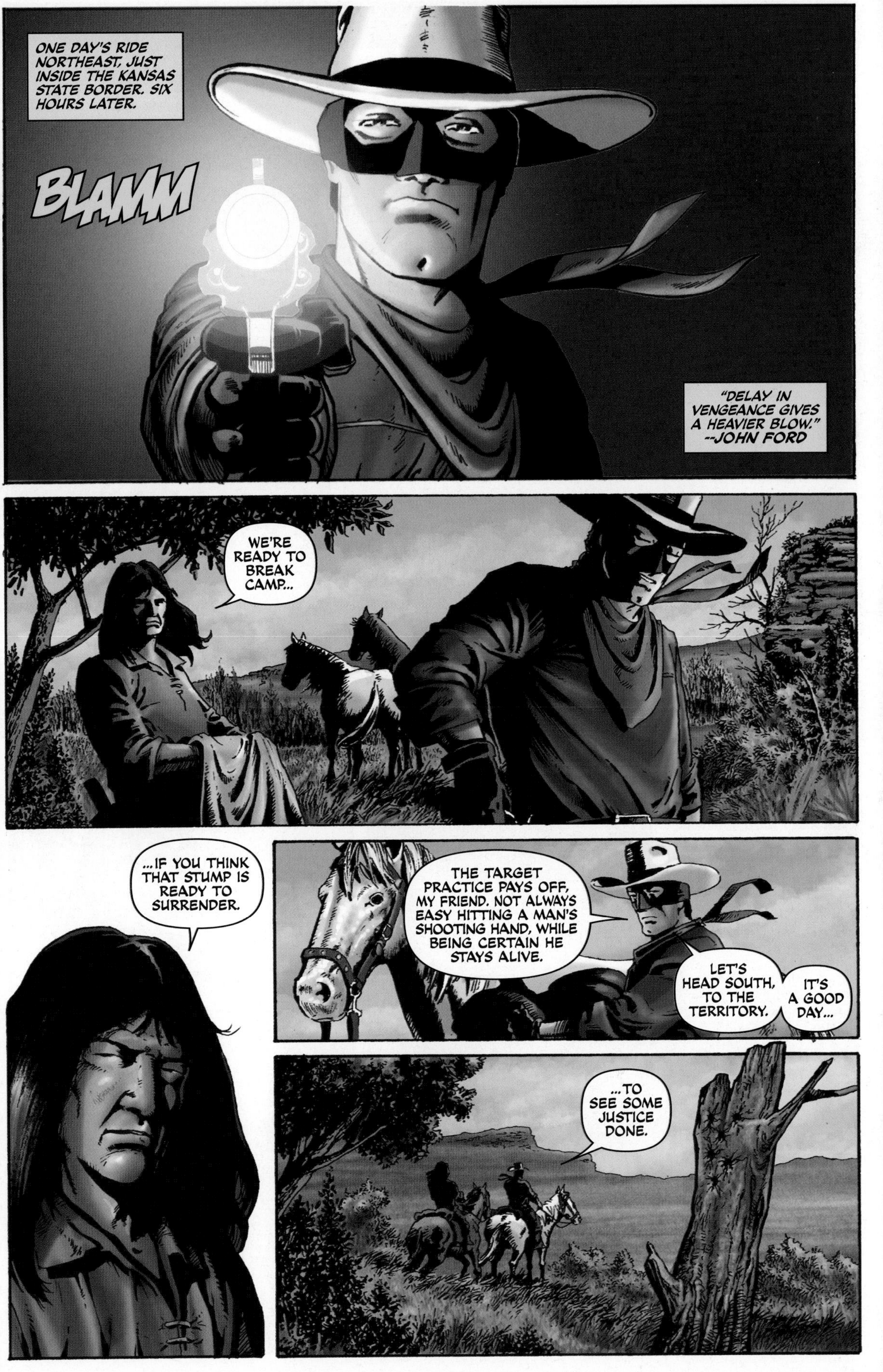
ONE DAY'S RIDE NORTHEAST, JUST INSIDE THE KANSAS STATE BORDER. SIX HOURS LATER.
BLAMM
"DELAY IN VENGEANCE GIVES A HEAVIER BLOW." --JOHN FORD
WE'RE READY TO BREAK CAMP...
...IF YOU THINK THAT STUMP IS READY TO SURRENDER.
THE TARGET PRACTICE PAYS OFF, MY FRIEND. NOT ALWAYS EASY HITTING A MAN'S SHOOTING HAND, WHILE BEING CERTAIN HE STAYS ALIVE.
LET'S HEAD SOUTH, TO THE TERRITORY.
IT'S A GOOD DAY...
...TO SEE SOME JUSTICE DONE.

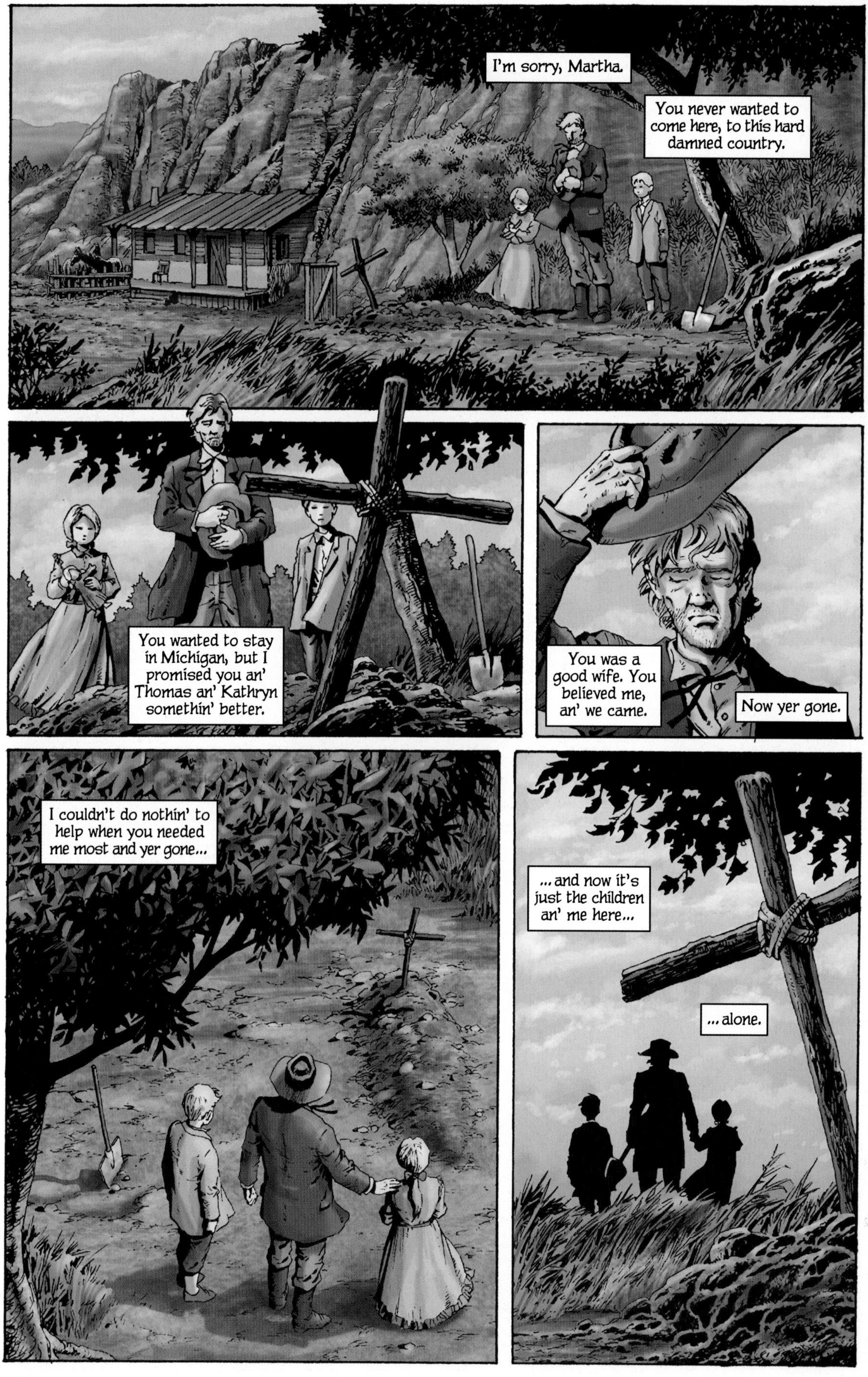
I'm sorry, Martha.
You never wanted to come here, to this hard damned country.
You wanted to stay in Michigan, but I promised you an' Thomas an' Kathryn somethin' better.
You was a good wife. You believed me, an' we came.
Now yer gone.
I couldn't do nothin' to help when you needed me most and yer gone...
...and now it's just the children an' me here...
...alone.

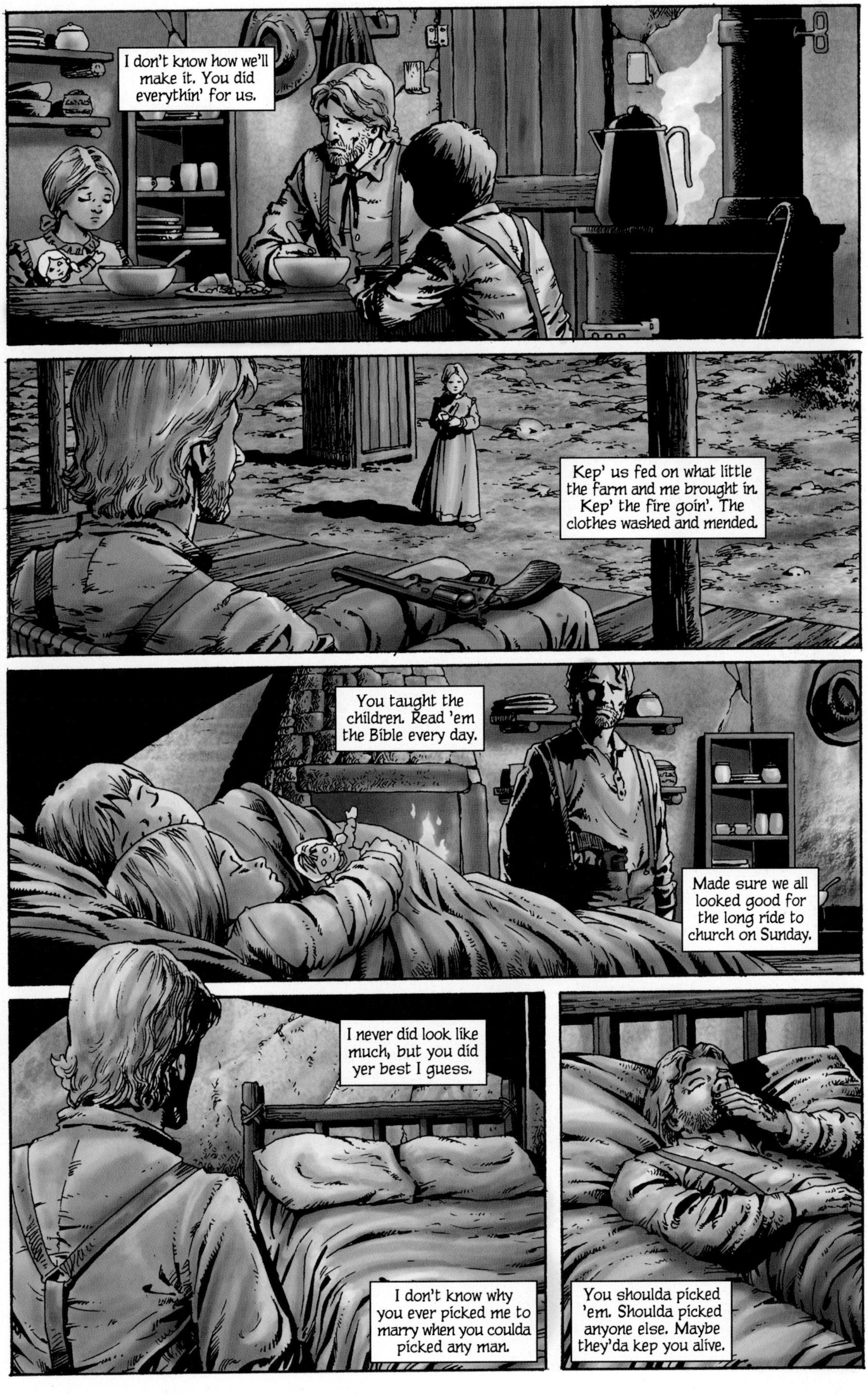
I don't know how we'll make it. You did everythin' for us.
Kep' us fed on what little the farm and me brought in. Kep' the fire goin'. The clothes washed and mended.
You taught the children. Read 'em the Bible every day.
Made sure we all looked good for the long ride to church on Sunday.
I never did look like much, but you did yer best I guess.
I don't know why you ever picked me to marry when you coulda picked any man.
You shoulda picked 'em. Shoulda picked anyone else. Maybe they'da kep you alive.

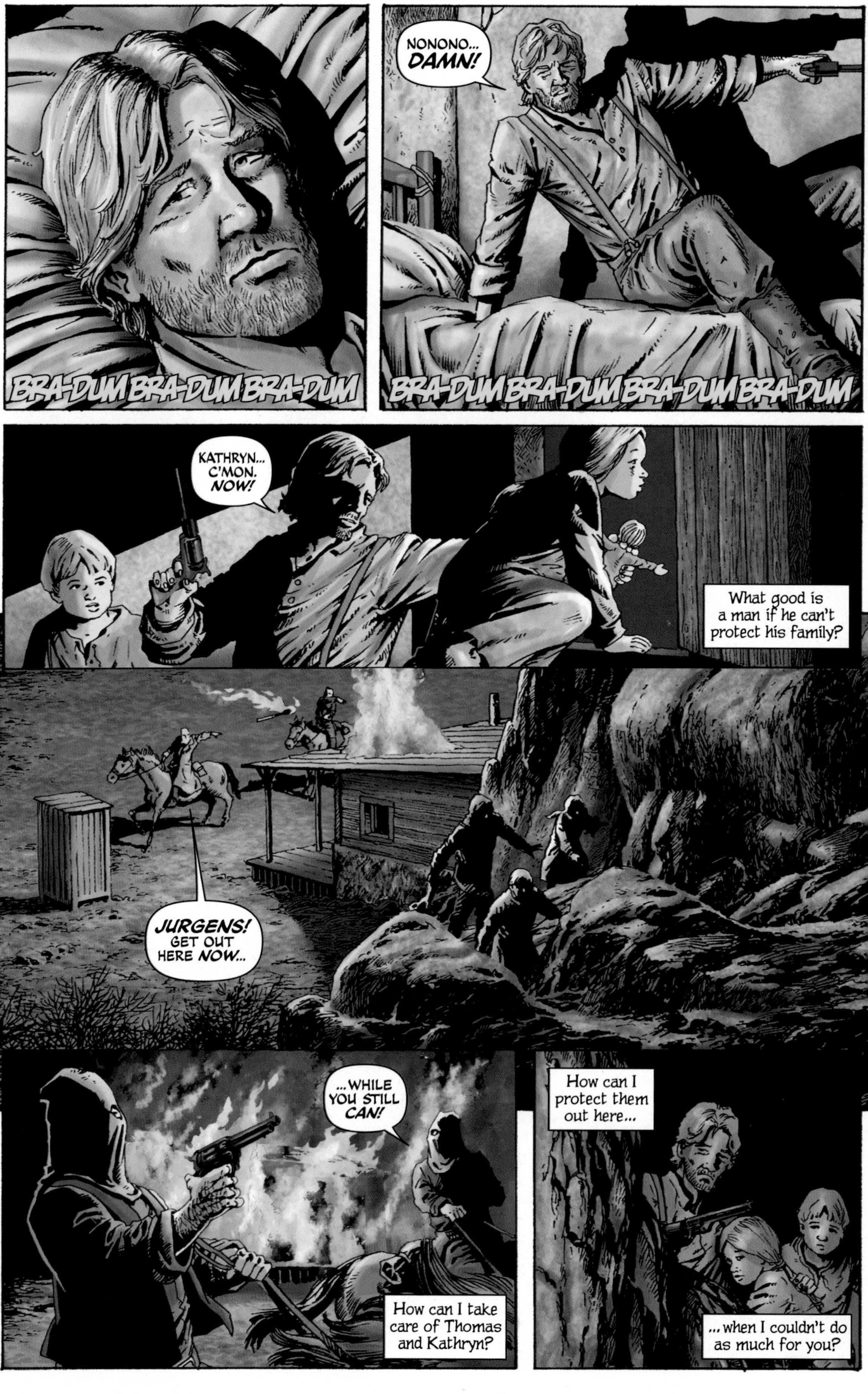
BRA-DUM BRA-DUM BRA-DUM
NONONO... DAMN!
BRA-DUM BRA-DUM BRA-DUM BRA-DUM
KATHRYN... C'MON. NOW!
What good is a man if he can't protect his family?
JURGENS! GET OUT HERE NOW...
...WHILE YOU STILL CAN!
How can I take care of Thomas and Kathryn?
How can I protect them out here...
...when I couldn't do as much for you?

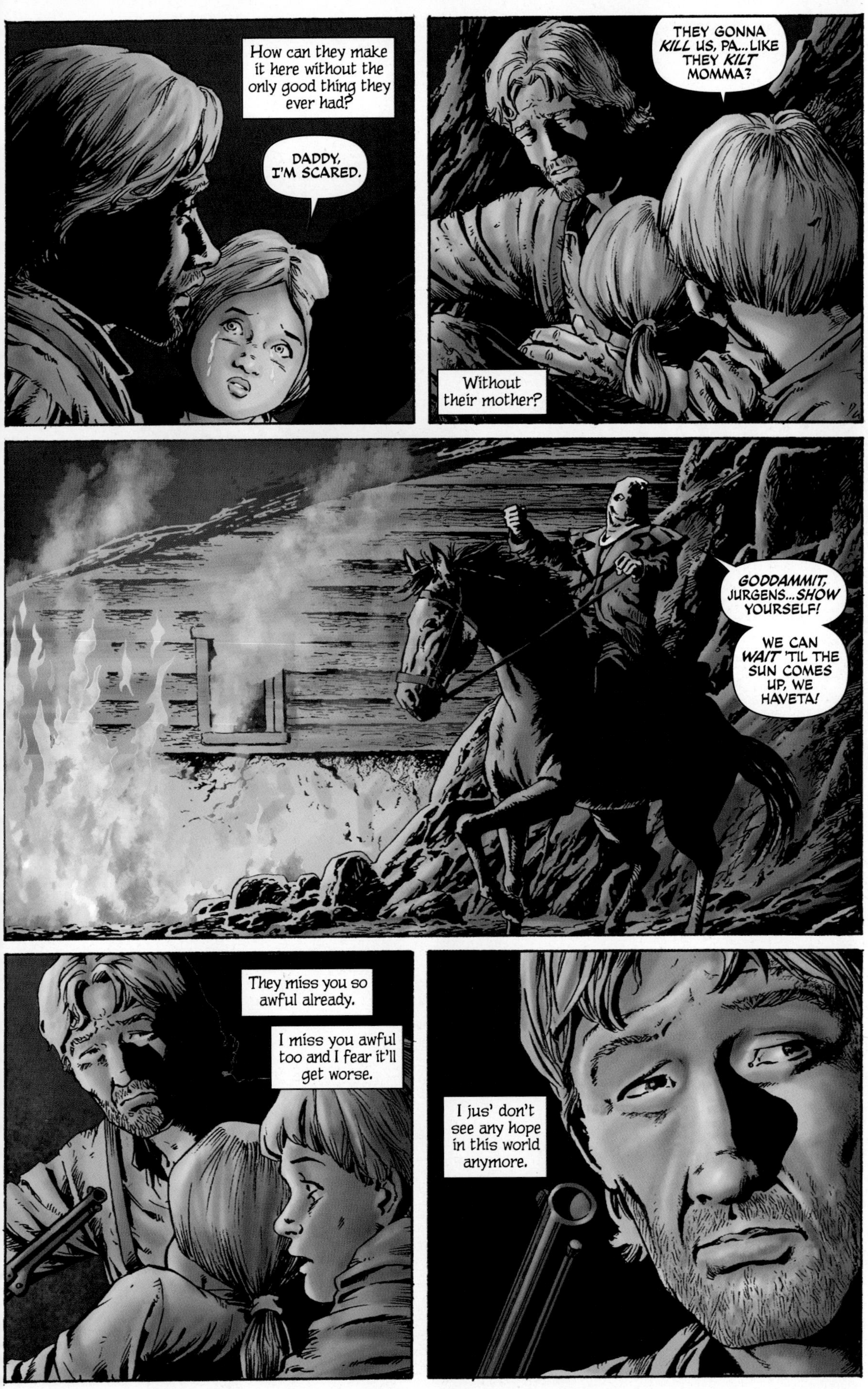

How can they make it here without the only good thing they ever had?
DADDY, I'M SCARED.
THEY GONNA KILL US, PA...LIKE THEY KILT MOMMA?
Without their mother?
GODDAMMIT, JURGENS...SHOW YOURSELF!
WE CAN WAIT 'TIL THE SUN COMES UP, WE HAVETA!
They miss you so awful already.
I miss you awful too and I fear it'll get worse.
I jus' don't see any hope in this world anymore.

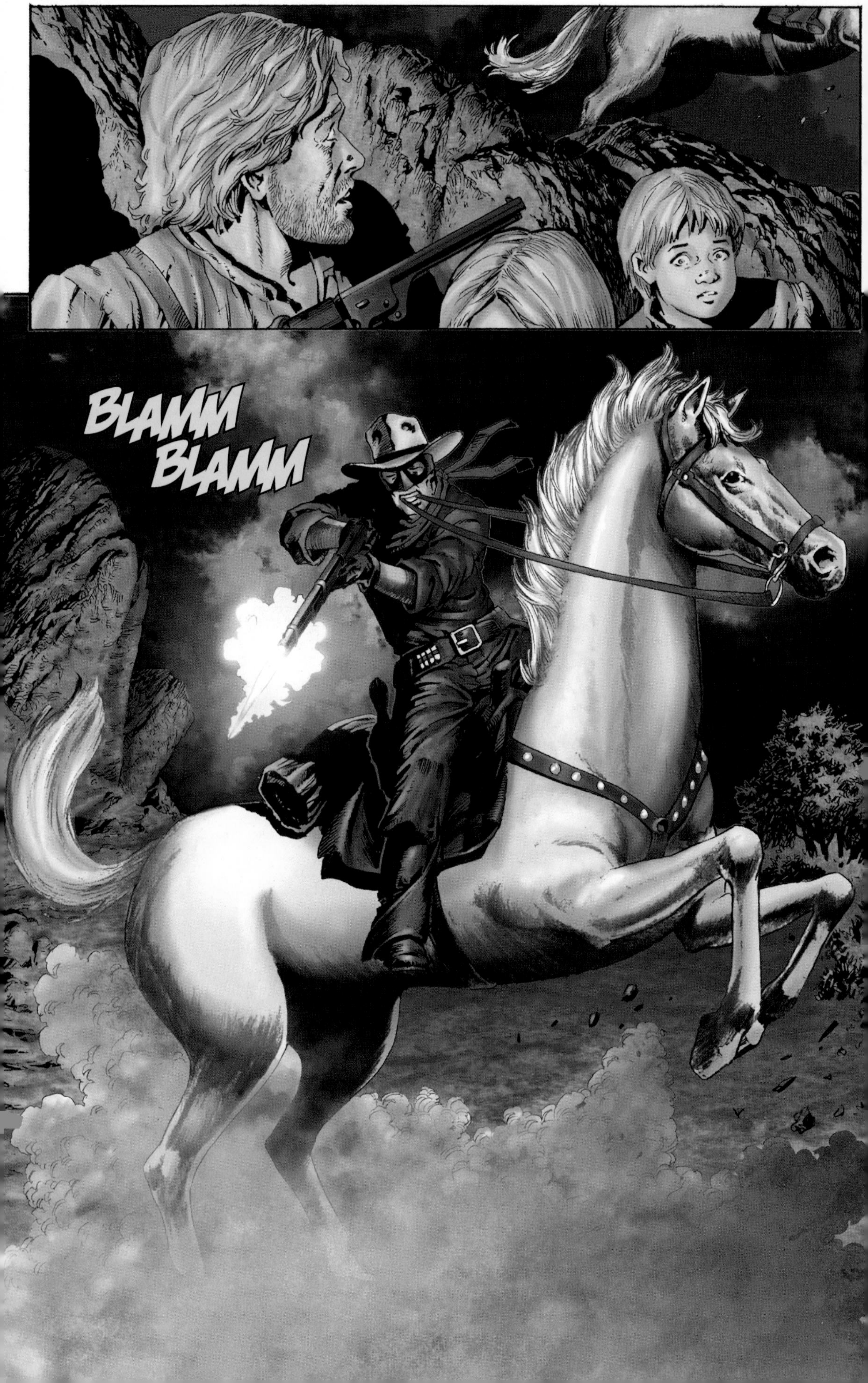
BLAMM
BLAMM

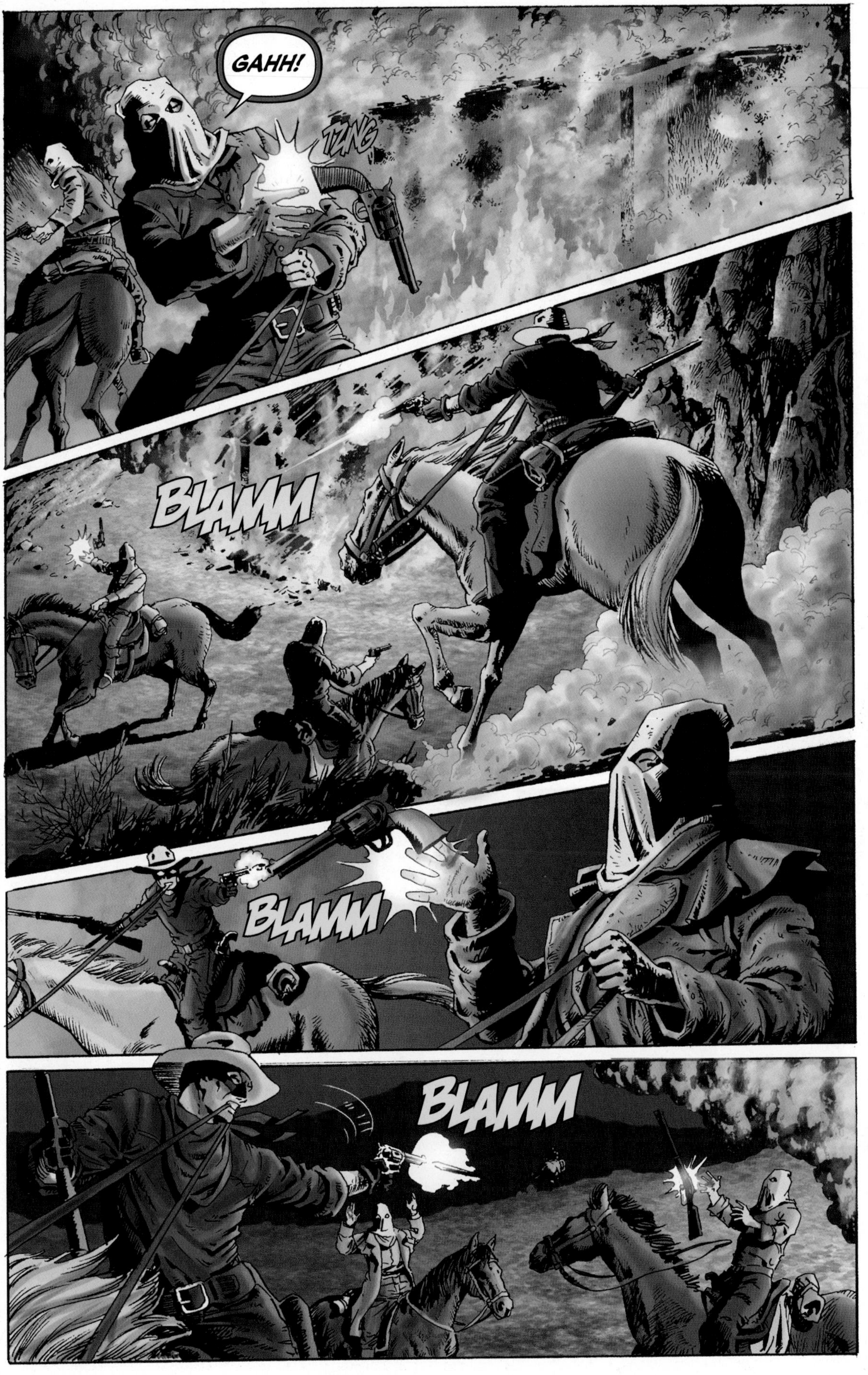
GAHH!
TZING
BLAMM
BLAMM
BLAMM

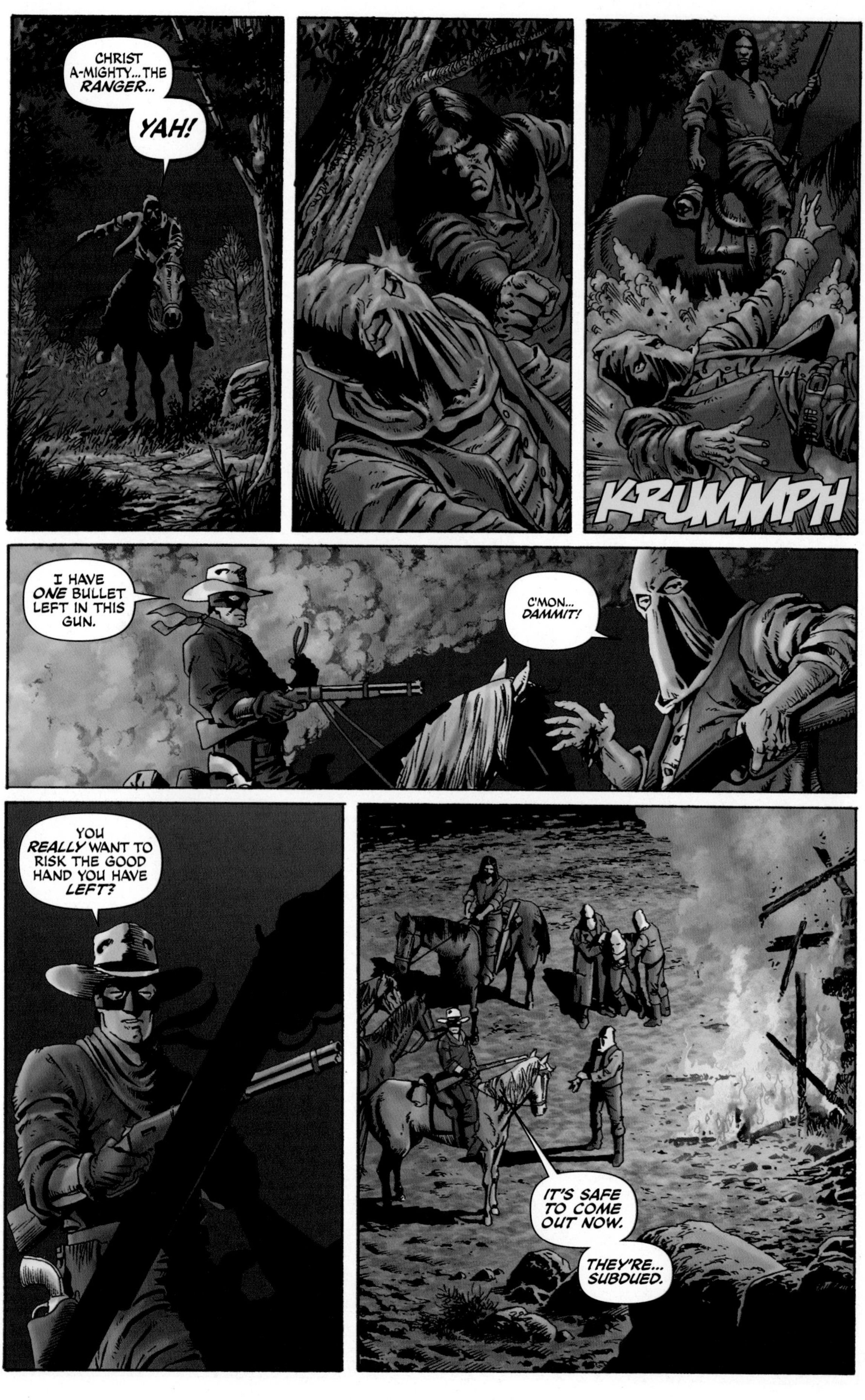
CHRIST A-MIGHTY... THE *RANGER*...
YAH!
KRUMMPH
I HAVE *ONE* BULLET LEFT IN THIS GUN.
C'MON... *DAMMIT!*
YOU *REALLY* WANT TO RISK THE GOOD HAND YOU HAVE *LEFT?*
IT'S SAFE TO COME OUT NOW.
THEY'RE... SUBDUED.

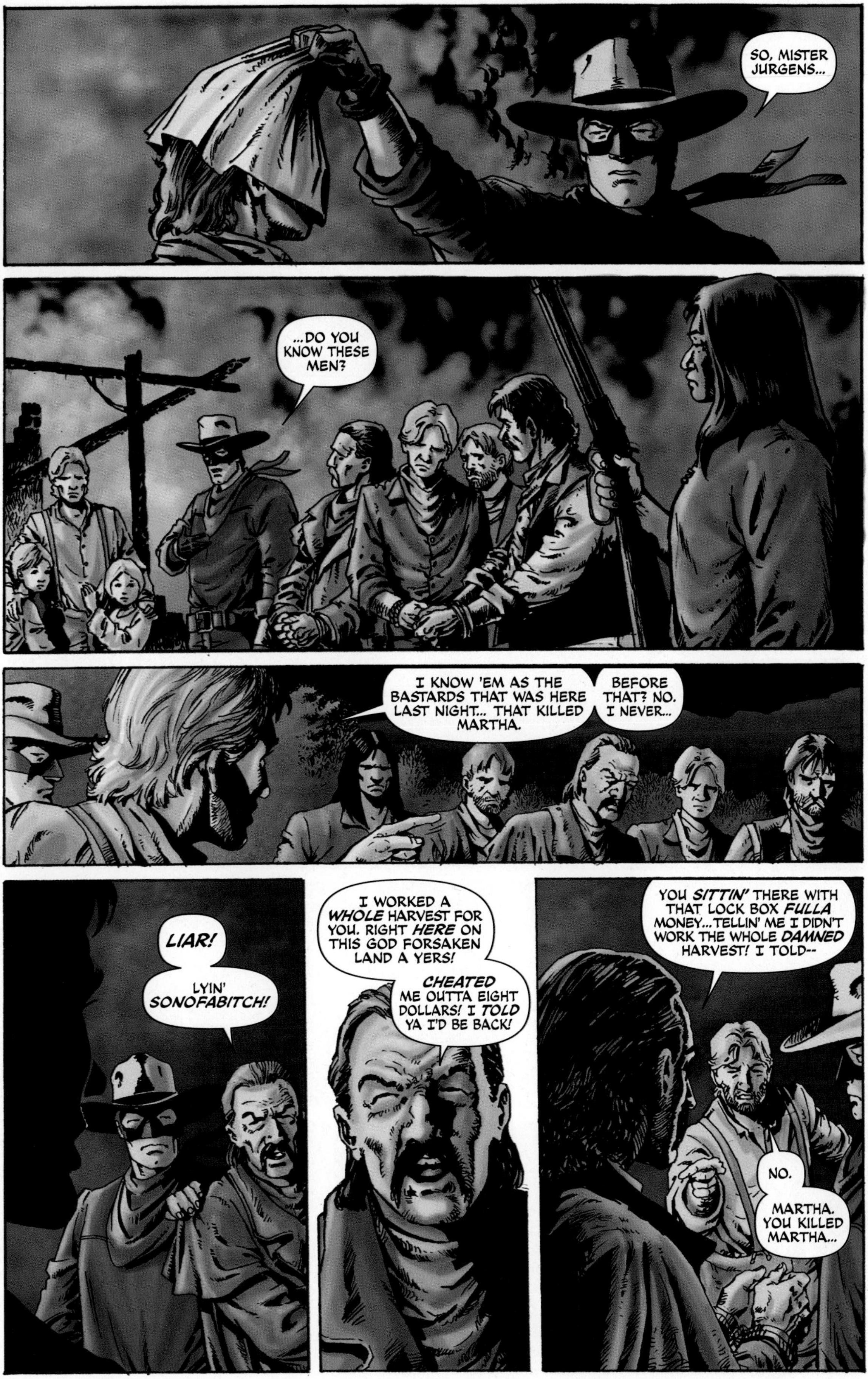
SO, MISTER JURGENS...
...DO YOU KNOW THESE MEN?
I KNOW 'EM AS THE BASTARDS THAT WAS HERE LAST NIGHT... THAT KILLED MARTHA.
BEFORE THAT? NO. I NEVER...
LIAR!
LYIN' SONOFABITCH!
I WORKED A WHOLE HARVEST FOR YOU. RIGHT HERE ON THIS GOD FORSAKEN LAND A YERS!
CHEATED ME OUTTA EIGHT DOLLARS! I TOLD YA I'D BE BACK!
YOU SITTIN' THERE WITH THAT LOCK BOX FULLA MONEY...TELLIN' ME I DIDN'T WORK THE WHOLE DAMNED HARVEST! I TOLD--
NO.
MARTHA. YOU KILLED MARTHA...

NOT *MY* FAULT SHE RAN OUT AN' GOT IN THE WAY! *WE* NEVER MADE HER DO THAT.

ALL FER EIGHT DOLLARS...

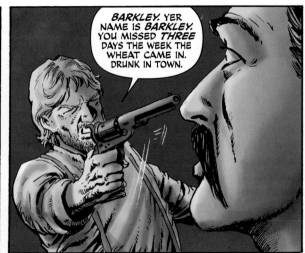

BARKLEY. YER NAME IS *BARKLEY.* YOU MISSED *THREE* DAYS THE WEEK THE WHEAT CAME IN. DRUNK IN TOWN.

TOO *LATE* TO HIRE 'NOTHER MAN. HAD TO LET SOME OF THE CROPS ROT. SHOULDN'TA PAID YOU AT ALL.

YOU CAME BACK OUT HERE LIKE THIS... FER EIGHT DOLLARS?

YOU *KILLED* MY MARTHA OVER EIGHT DOLLARS?

NATHAN. NOT LIKE *THIS.*

THE MAN WILL GET JUSTICE, BUT NOT LIKE *THIS.* YOUR *CHILDREN...*

YER A DAMNED *CHEAT,* JURGENS! YOU *DESERVE--* HUKKK!

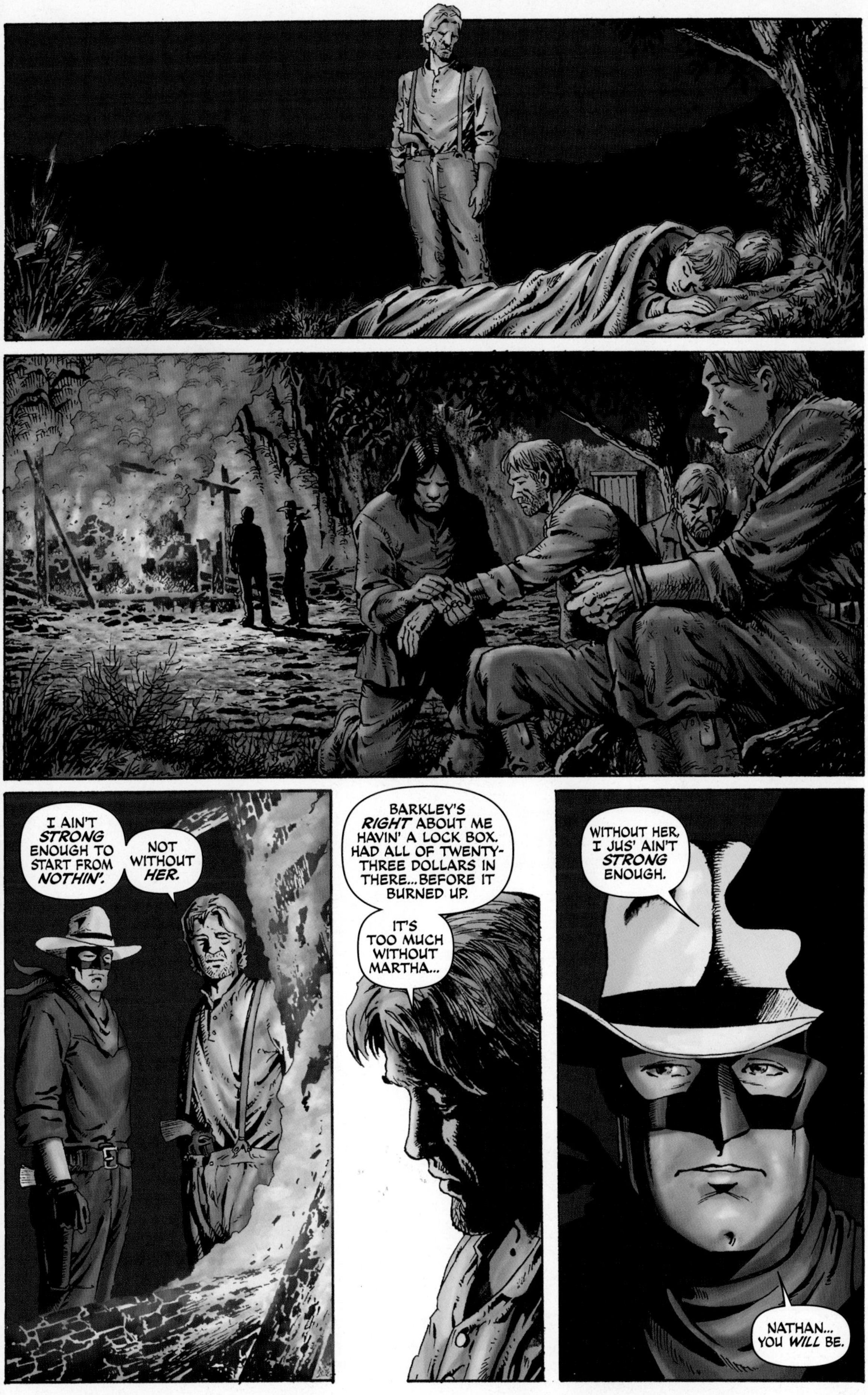
I AIN'T STRONG ENOUGH TO START FROM NOTHIN'.
NOT WITHOUT HER.
BARKLEY'S RIGHT ABOUT ME HAVIN' A LOCK BOX. HAD ALL OF TWENTY-THREE DOLLARS IN THERE...BEFORE IT BURNED UP.
IT'S TOO MUCH WITHOUT MARTHA...
WITHOUT HER, I JUS' AIN'T STRONG ENOUGH.
NATHAN... YOU WILL BE.

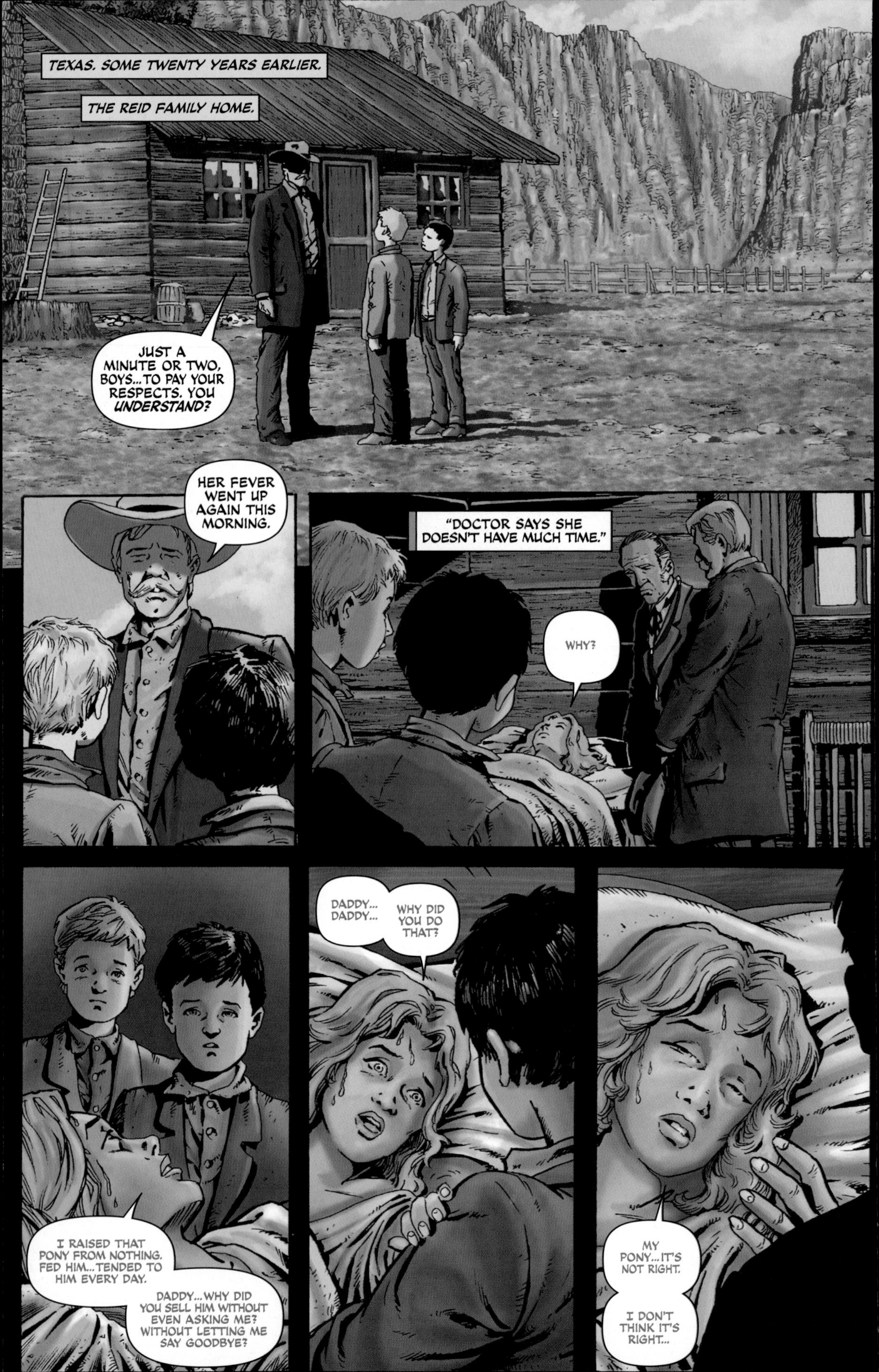
TEXAS. SOME TWENTY YEARS EARLIER.
THE REID FAMILY HOME.
JUST A MINUTE OR TWO, BOYS...TO PAY YOUR RESPECTS. YOU *UNDERSTAND?*
HER FEVER WENT UP AGAIN THIS MORNING.
"DOCTOR SAYS SHE DOESN'T HAVE MUCH TIME."
WHY?
I RAISED THAT PONY FROM NOTHING. FED HIM...TENDED TO HIM EVERY DAY.
DADDY...WHY DID YOU SELL HIM WITHOUT EVEN ASKING ME? WITHOUT LETTING ME SAY GOODBYE?
DADDY... DADDY...
WHY DID YOU DO THAT?
MY PONY...IT'S NOT RIGHT.
I DON'T THINK IT'S RIGHT...

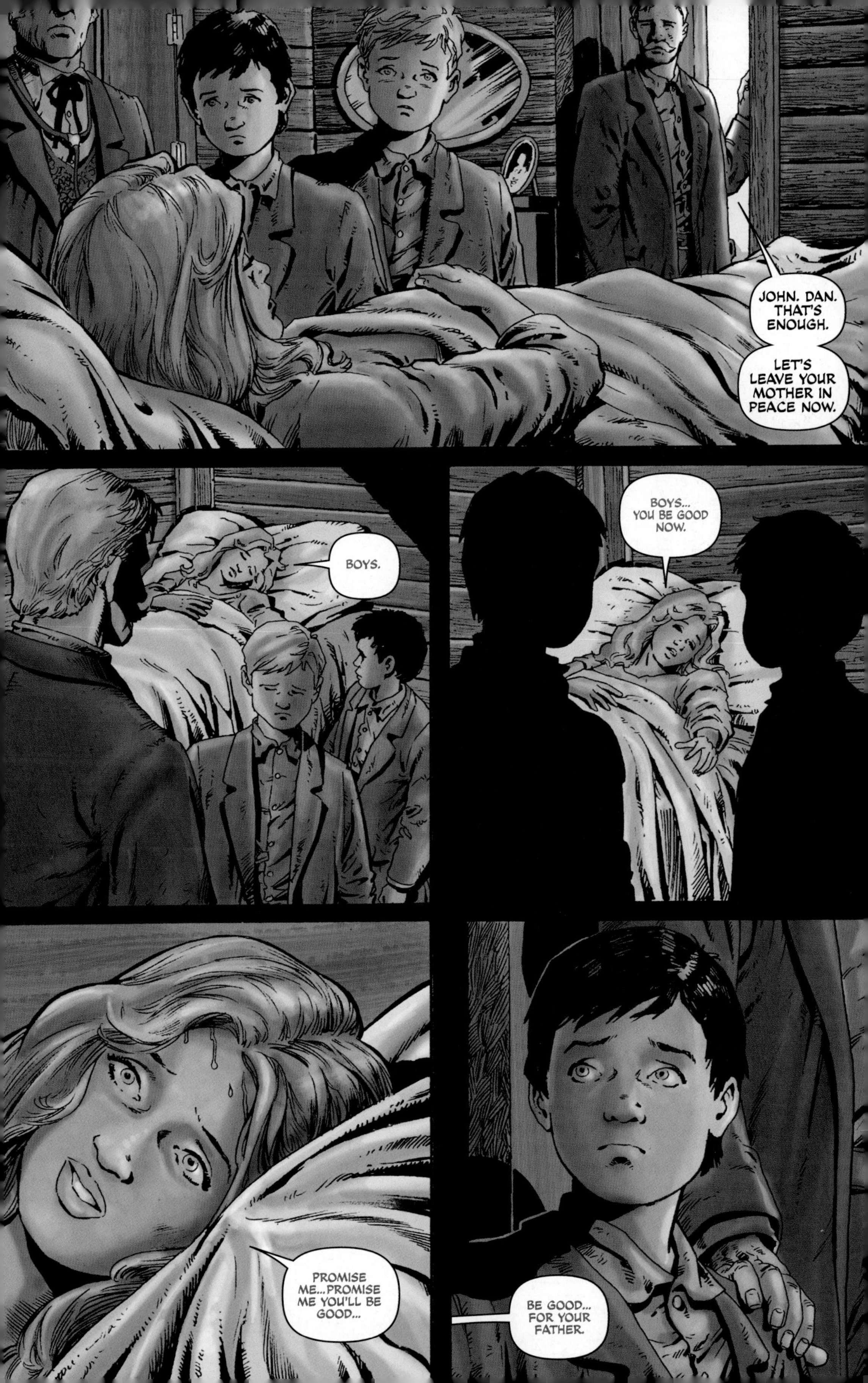
JOHN. DAN. THAT'S ENOUGH.
LET'S LEAVE YOUR MOTHER IN PEACE NOW.
BOYS.
BOYS... YOU BE GOOD NOW.
PROMISE ME...PROMISE ME YOU'LL BE GOOD...
BE GOOD... FOR YOUR FATHER.

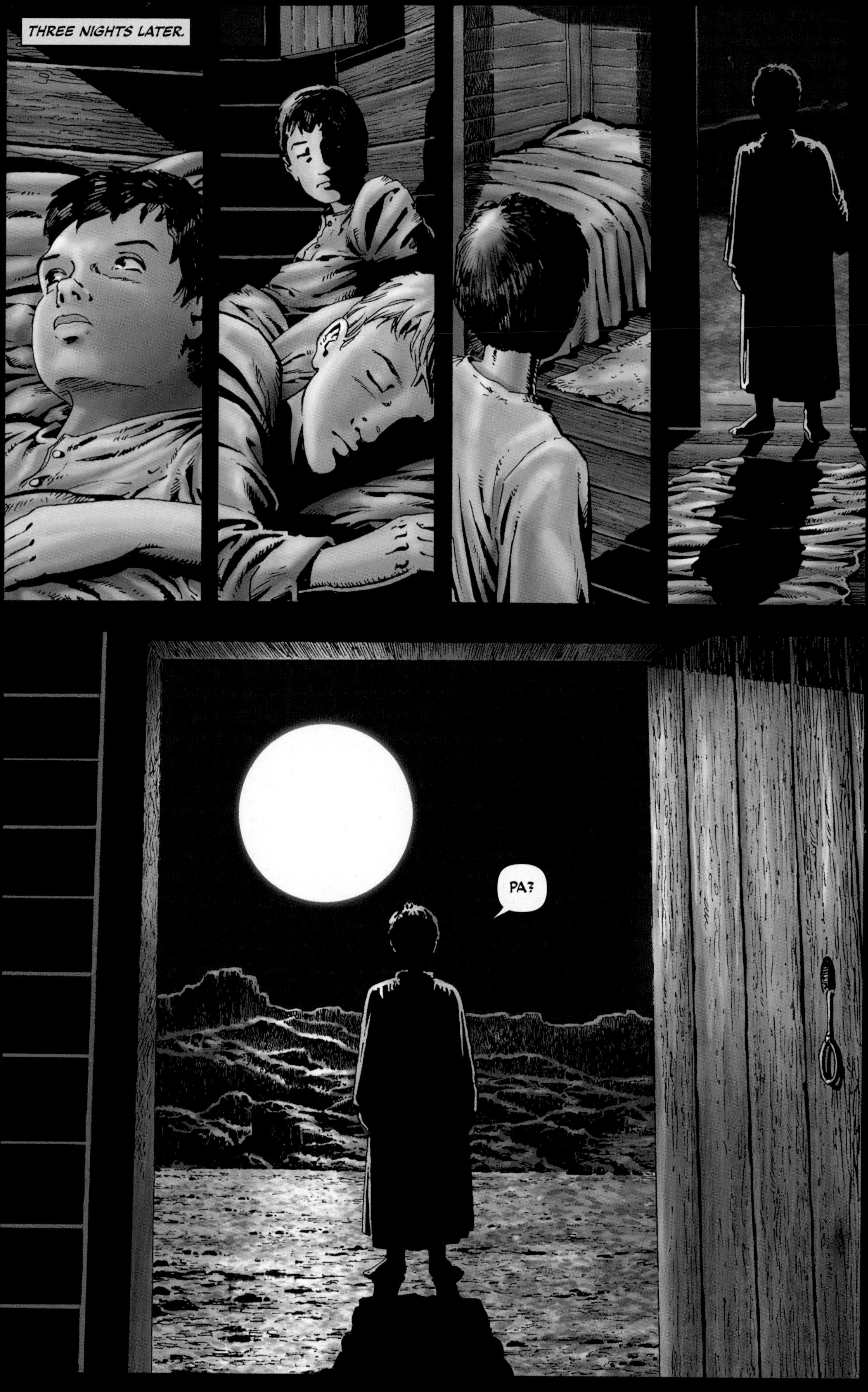
THREE NIGHTS LATER.
PA?

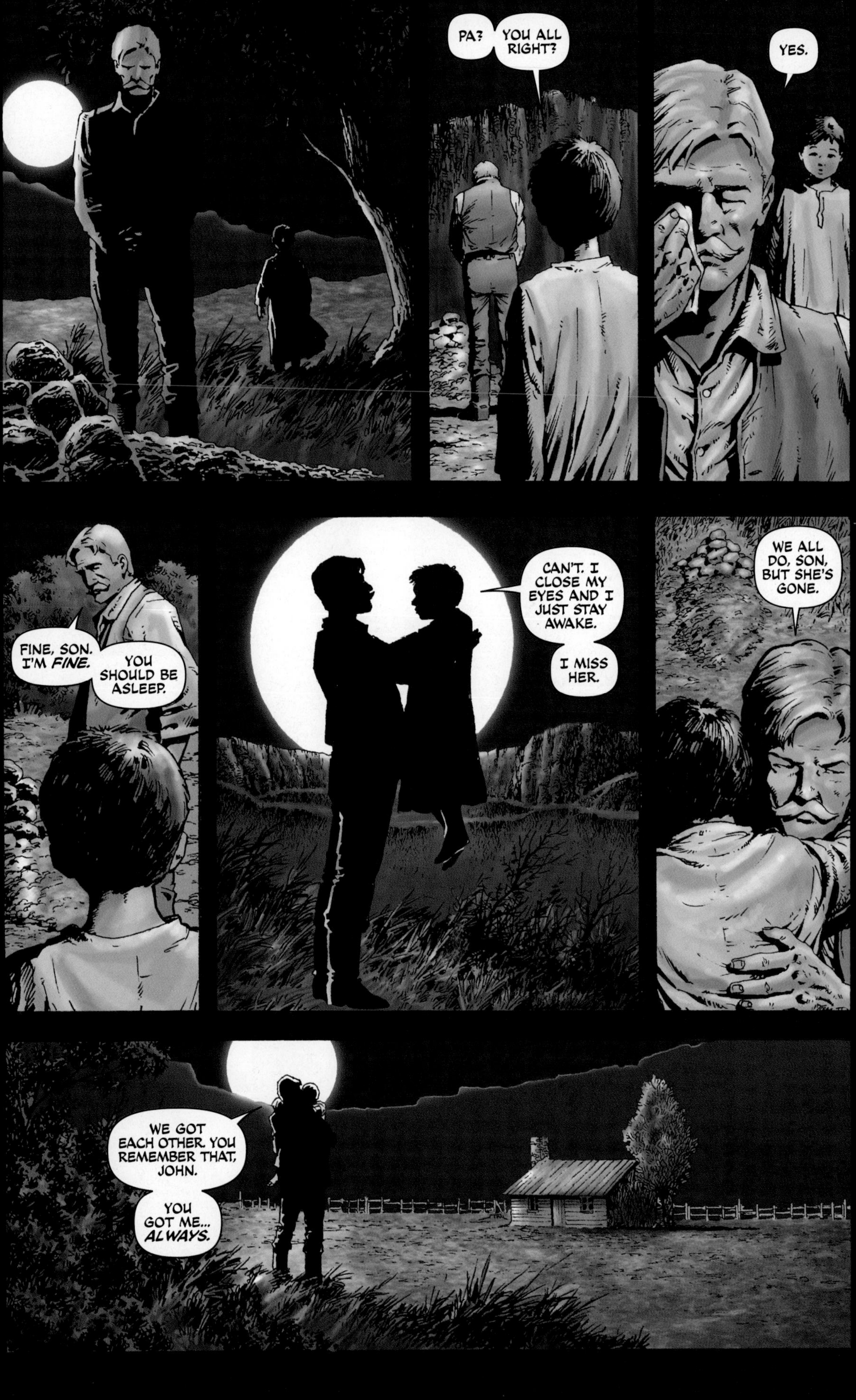
PA?
YOU ALL RIGHT?
YES.
FINE, SON. I'M *FINE.*
YOU SHOULD BE ASLEEP.
CAN'T. I CLOSE MY EYES AND I JUST STAY AWAKE.
I MISS HER.
WE ALL DO, SON, BUT SHE'S GONE.
WE GOT EACH OTHER. YOU REMEMBER THAT, JOHN.
YOU GOT ME... *ALWAYS.*

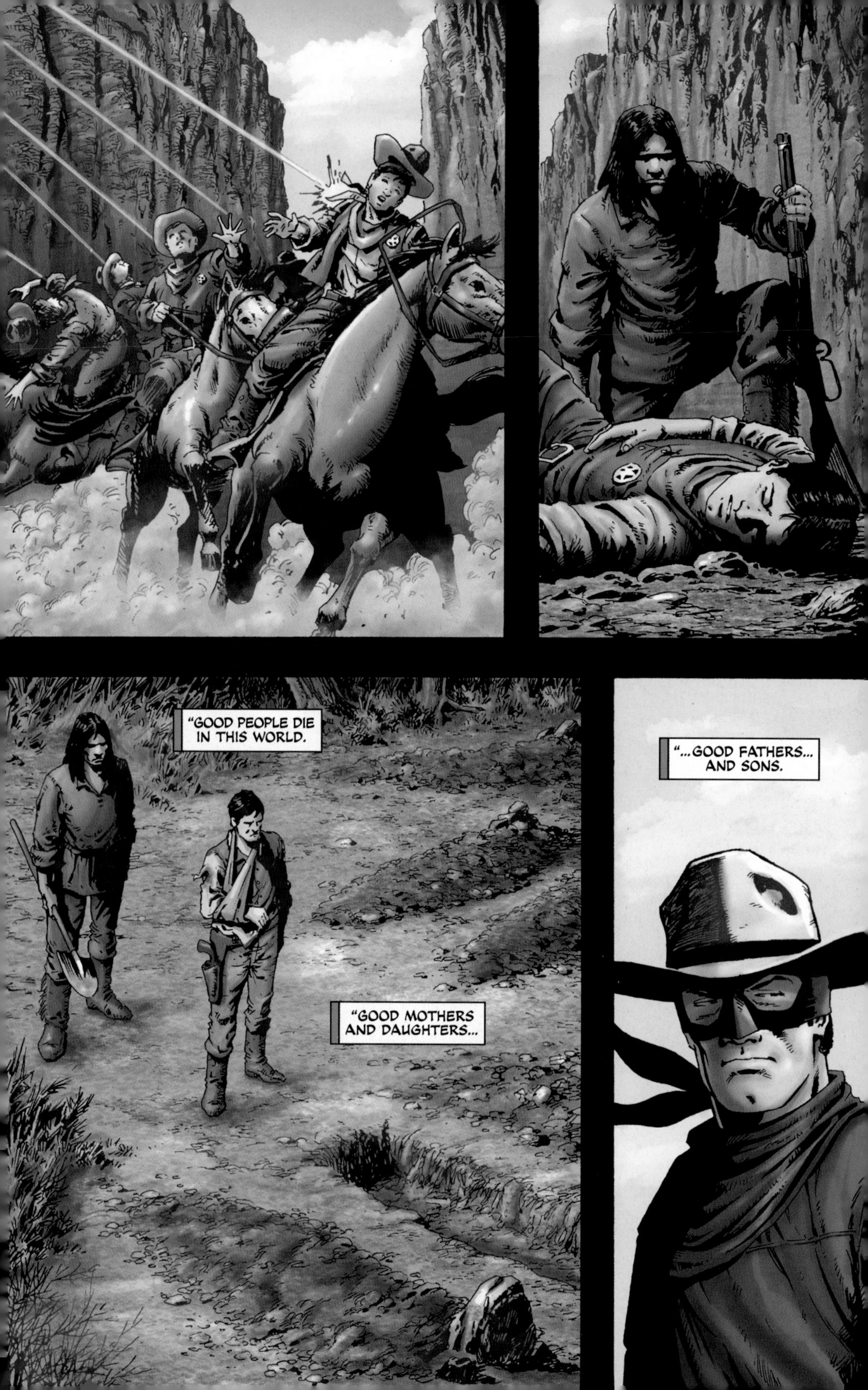
"GOOD PEOPLE DIE IN THIS WORLD.
"GOOD MOTHERS AND DAUGHTERS...
"...GOOD FATHERS... AND SONS.

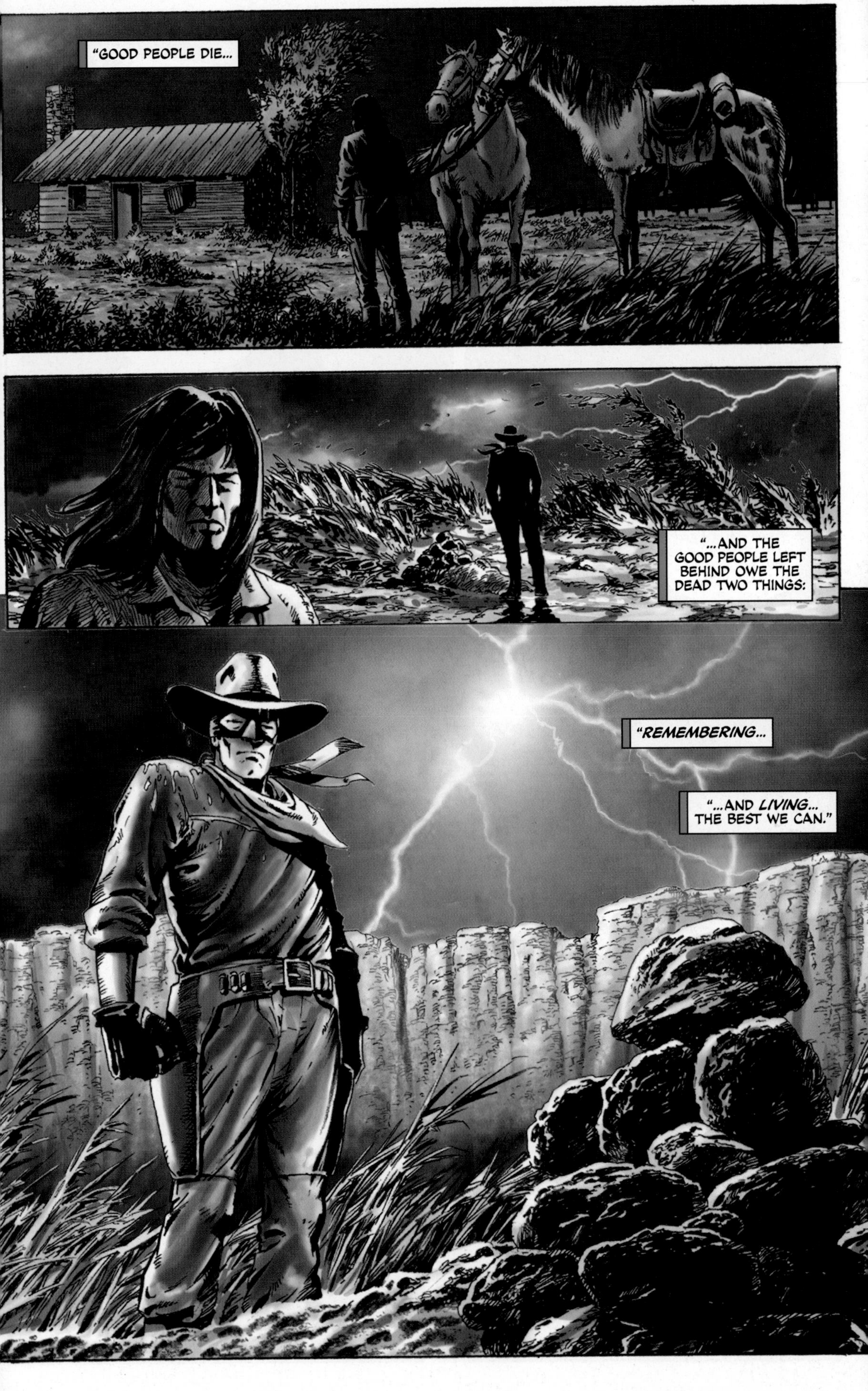
"GOOD PEOPLE DIE...
"...AND THE GOOD PEOPLE LEFT BEHIND OWE THE DEAD TWO THINGS:
"REMEMBERING...
"...AND LIVING... THE BEST WE CAN."

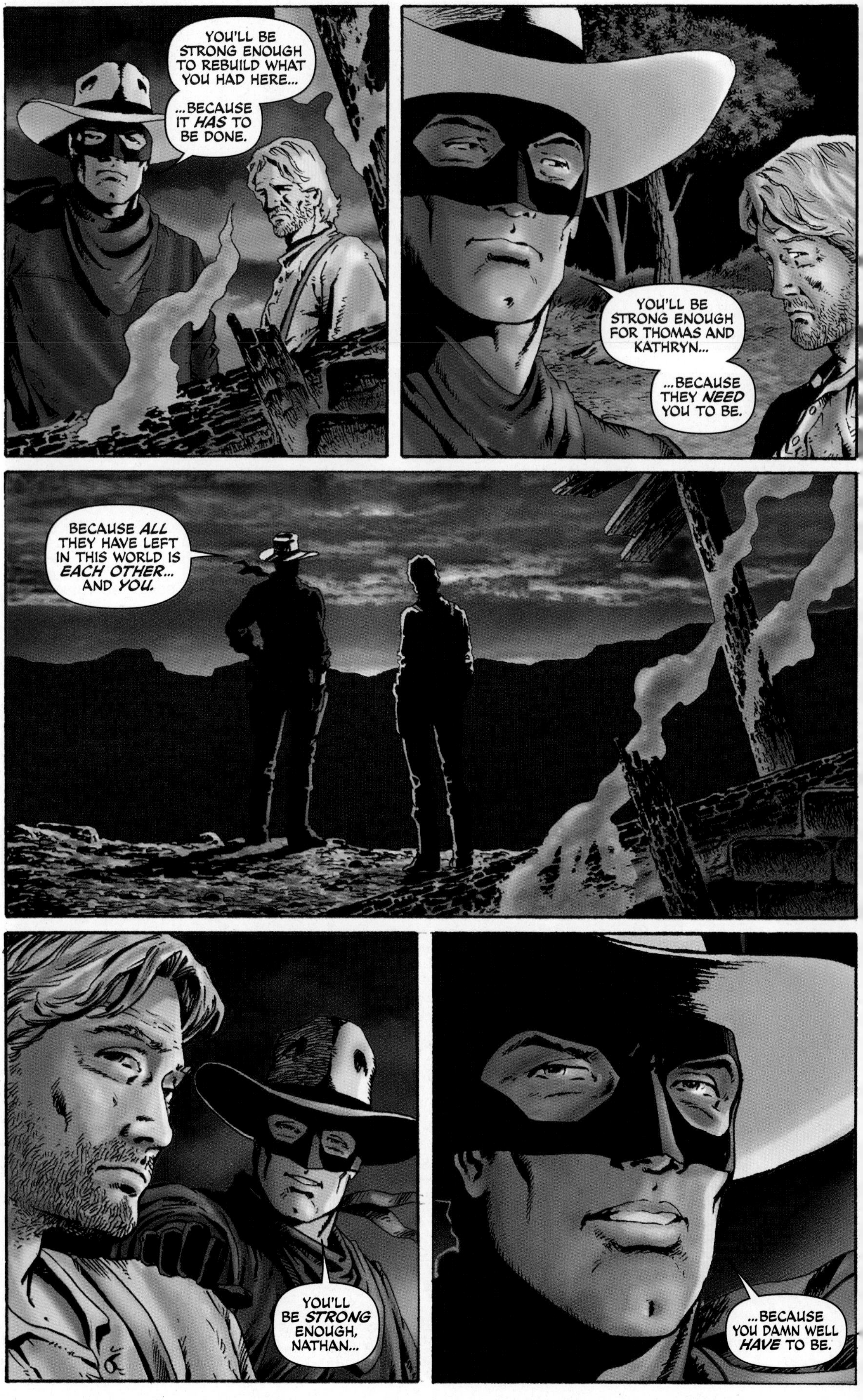
YOU'LL BE STRONG ENOUGH TO REBUILD WHAT YOU HAD HERE...
...BECAUSE IT HAS TO BE DONE.
YOU'LL BE STRONG ENOUGH FOR THOMAS AND KATHRYN...
...BECAUSE THEY NEED YOU TO BE.
BECAUSE ALL THEY HAVE LEFT IN THIS WORLD IS EACH OTHER... AND YOU.
YOU'LL BE STRONG ENOUGH, NATHAN...
...BECAUSE YOU DAMN WELL HAVE TO BE.

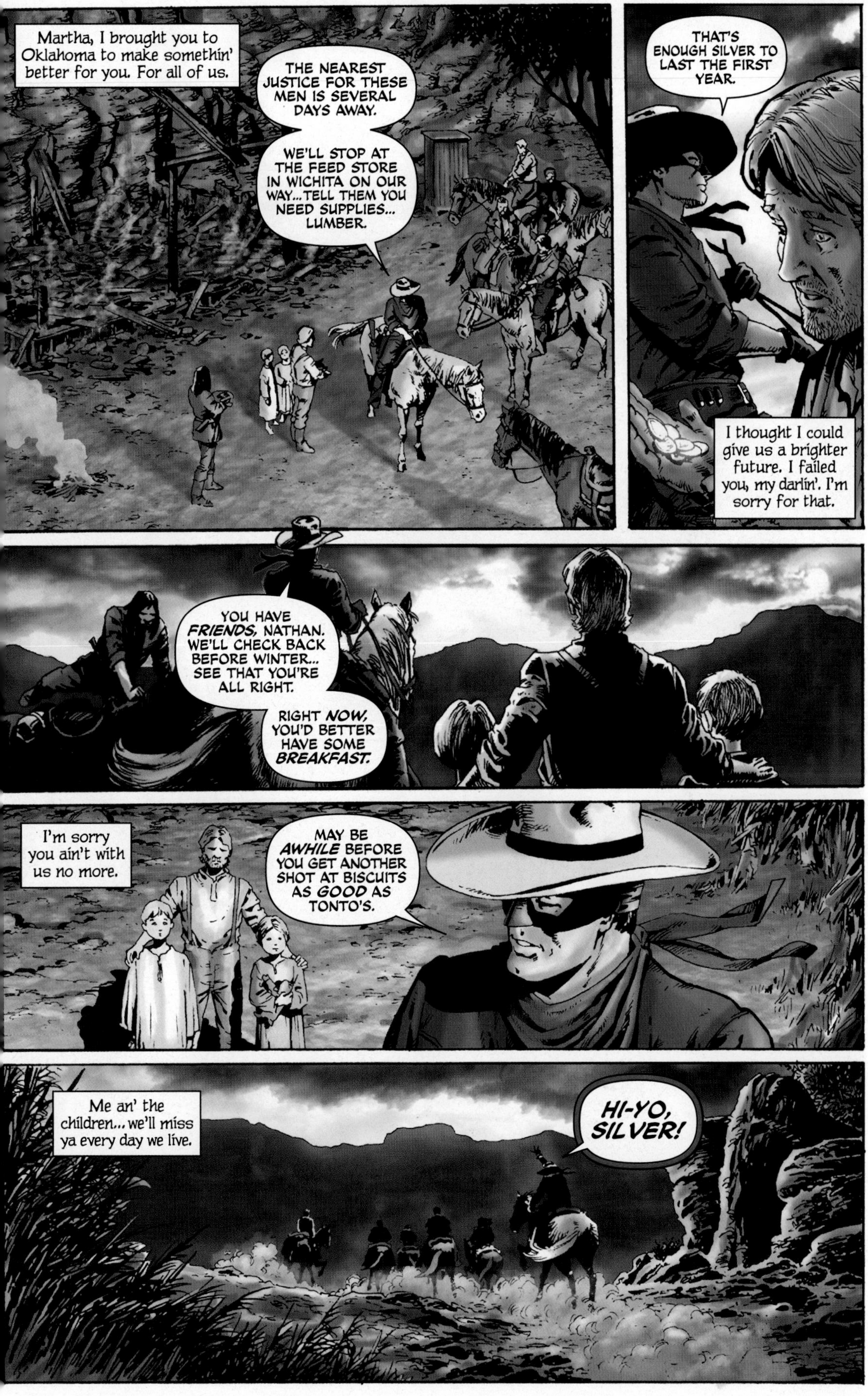
Martha, I brought you to Oklahoma to make somethin' better for you. For all of us.
THE NEAREST JUSTICE FOR THESE MEN IS SEVERAL DAYS AWAY.
WE'LL STOP AT THE FEED STORE IN WICHITA ON OUR WAY...TELL THEM YOU NEED SUPPLIES... LUMBER.
THAT'S ENOUGH SILVER TO LAST THE FIRST YEAR.
I thought I could give us a brighter future. I failed you, my darlin'. I'm sorry for that.
YOU HAVE FRIENDS, NATHAN. WE'LL CHECK BACK BEFORE WINTER... SEE THAT YOU'RE ALL RIGHT.
RIGHT NOW, YOU'D BETTER HAVE SOME BREAKFAST.
I'm sorry you ain't with us no more.
MAY BE AWHILE BEFORE YOU GET ANOTHER SHOT AT BISCUITS AS GOOD AS TONTO'S.
Me an' the children... we'll miss ya every day we live.
HI-YO, SILVER!

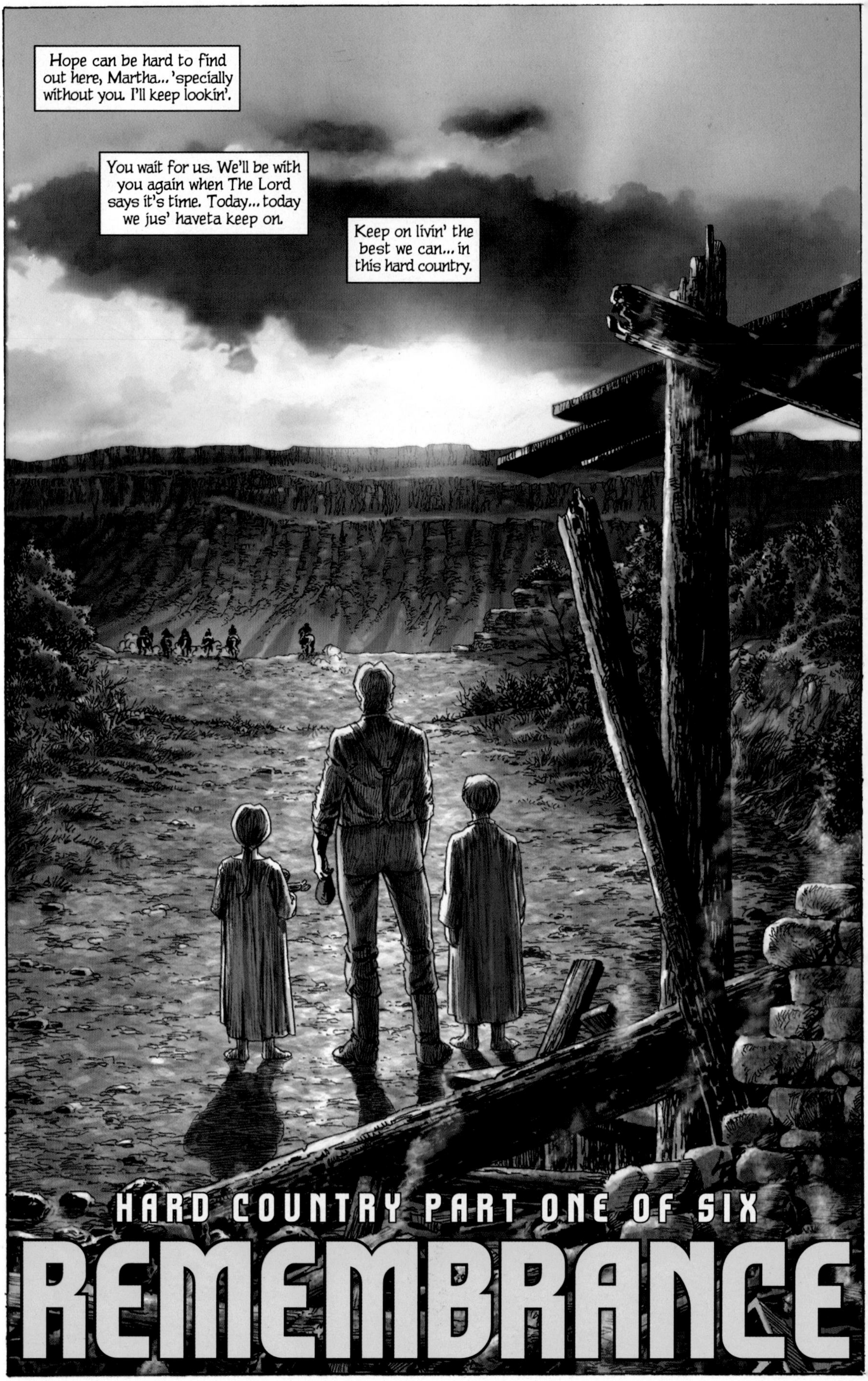
Hope can be hard to find out here, Martha... 'specially without you. I'll keep lookin'.
You wait for us. We'll be with you again when The Lord says it's time. Today... today we jus' haveta keep on.
Keep on livin' the best we can... in this hard country.
HARD COUNTRY PART ONE OF SIX
REMEMBRANCE

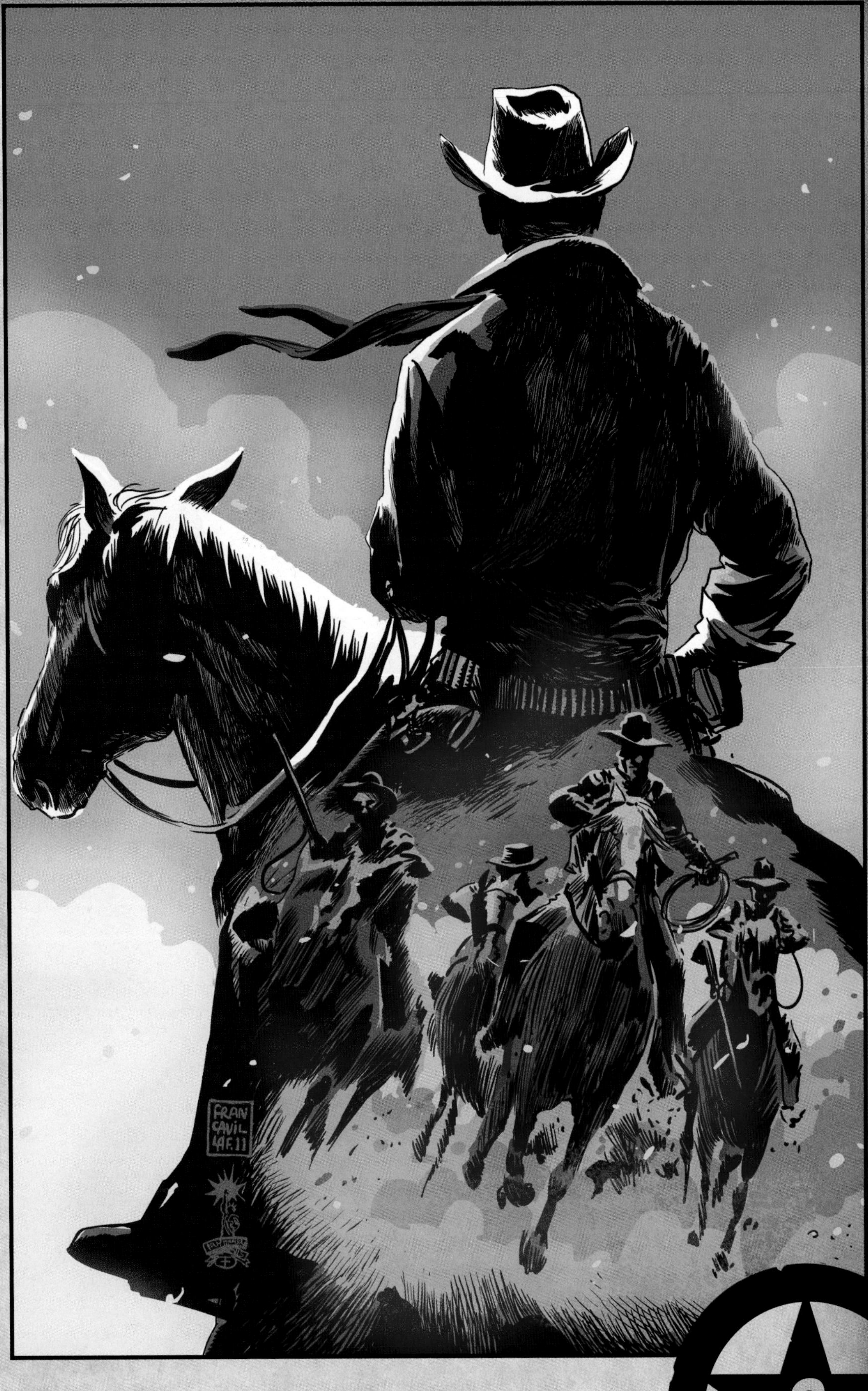
FRAN CAVIL LA F.11

COVER BY FRANCESCO FRANCAVILLA

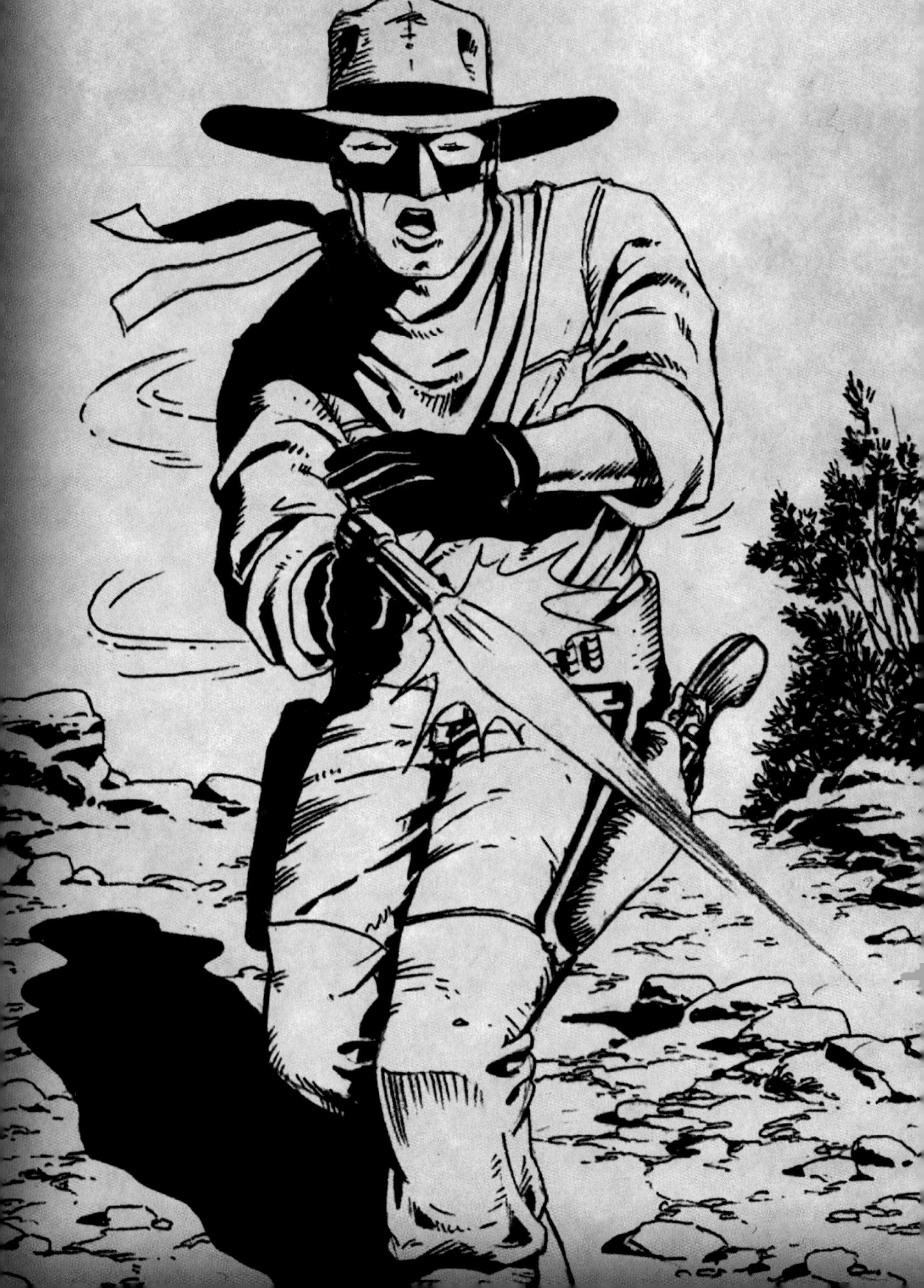

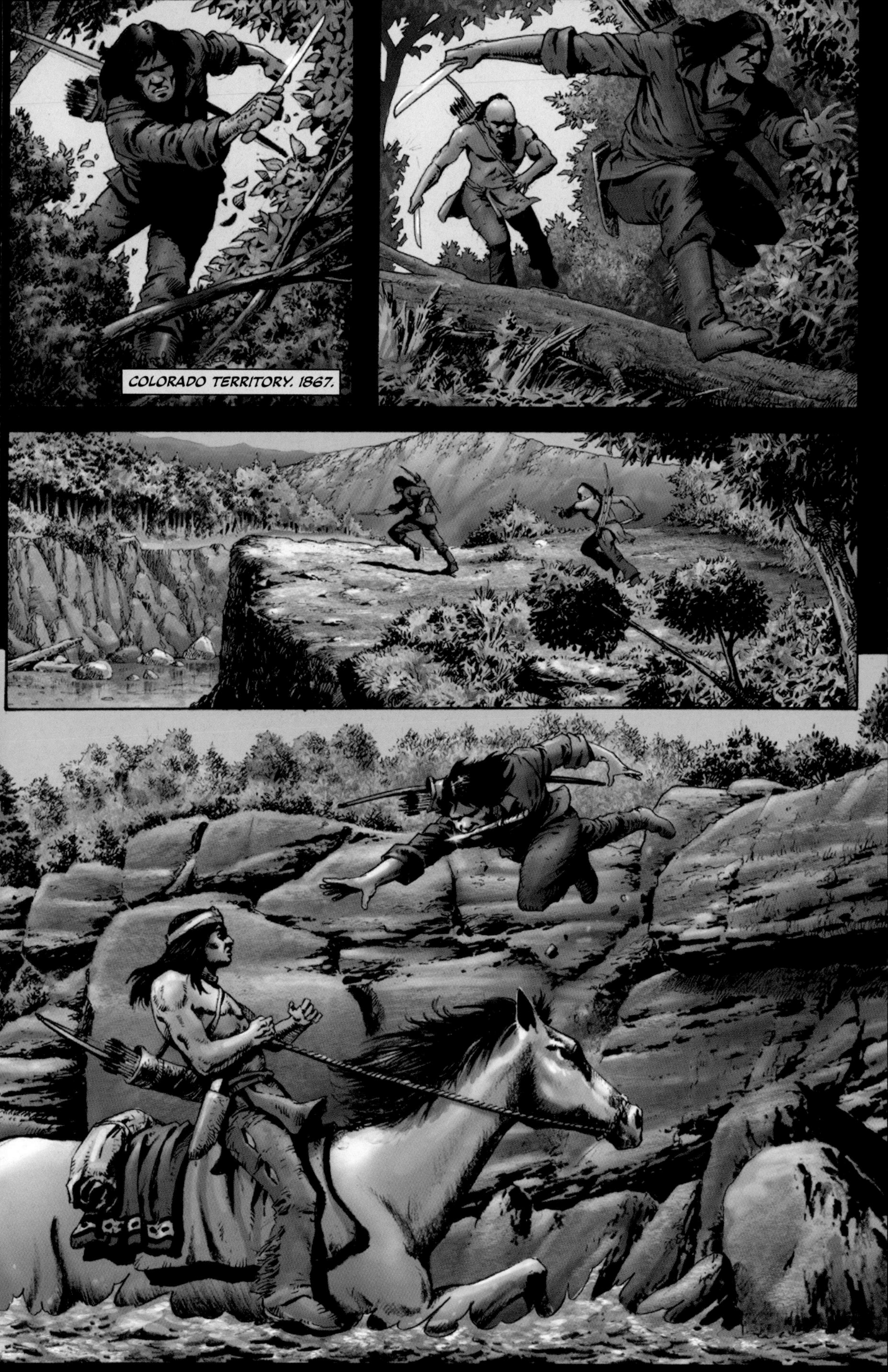
COLORADO TERRITORY. 1867.

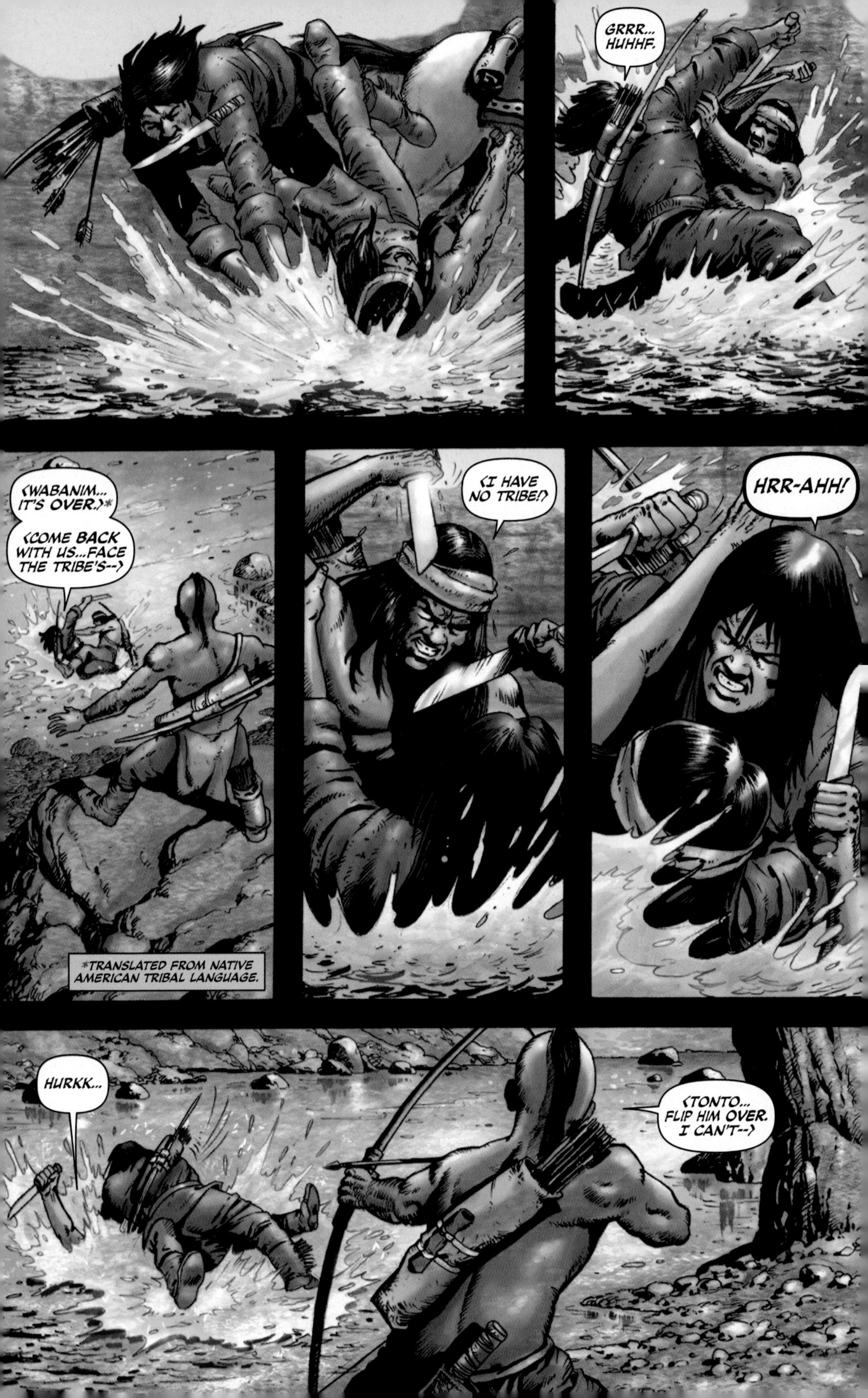
GRRR... HUHHF.
<WABANIM... IT'S OVER.>*
<COME BACK WITH US...FACE THE TRIBE'S-->
*TRANSLATED FROM NATIVE AMERICAN TRIBAL LANGUAGE.
<I HAVE NO TRIBE!>
HRR-AHH!
HURKK...
<TONTO... FLIP HIM OVER. I CAN'T-->

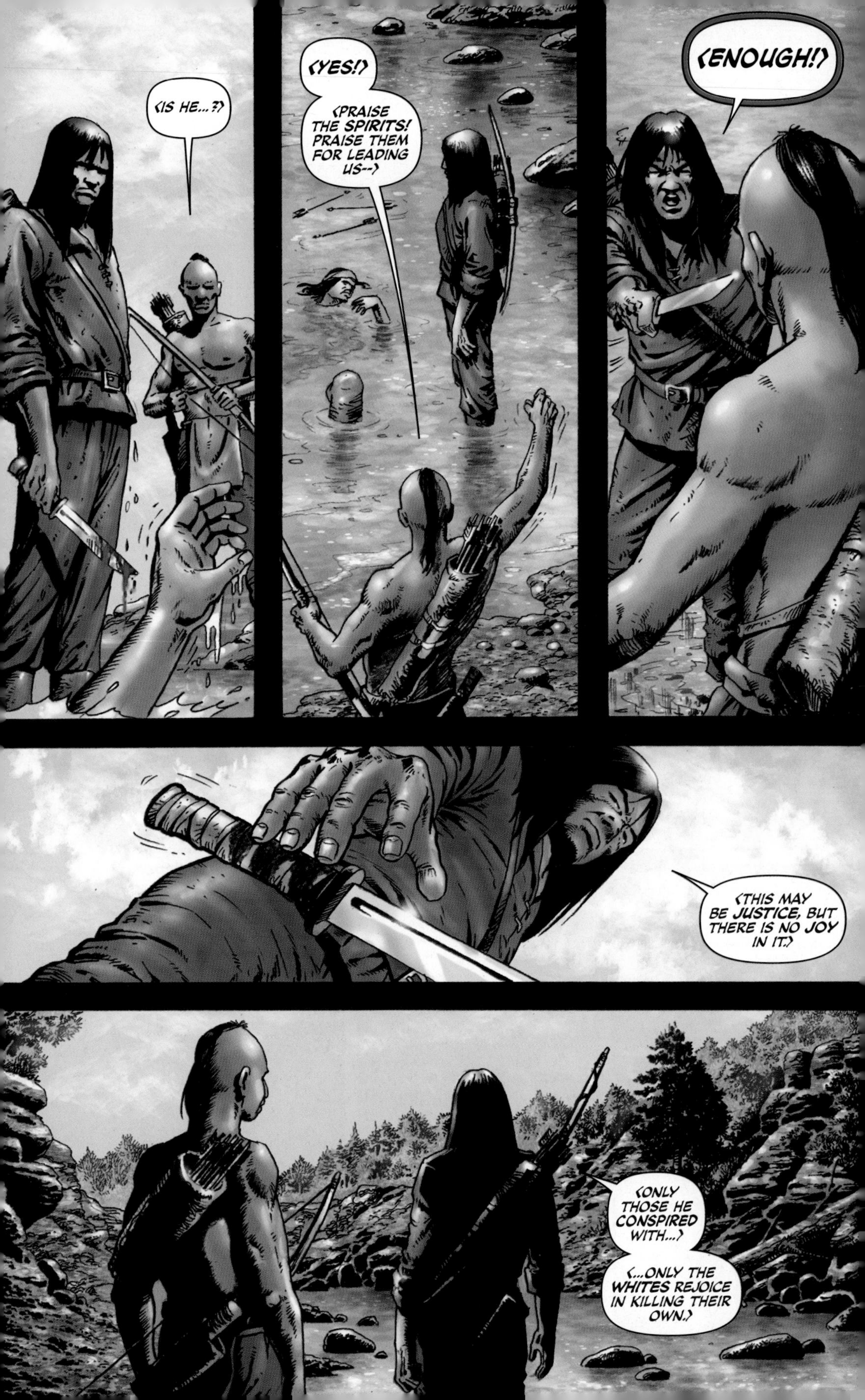

<IS HE...?>
<YES!>
<PRAISE THE SPIRITS! PRAISE THEM FOR LEADING US-->
<ENOUGH!>
<THIS MAY BE JUSTICE, BUT THERE IS NO JOY IN IT.>
<ONLY THOSE HE CONSPIRED WITH...>
<...ONLY THE WHITES REJOICE IN KILLING THEIR OWN.>

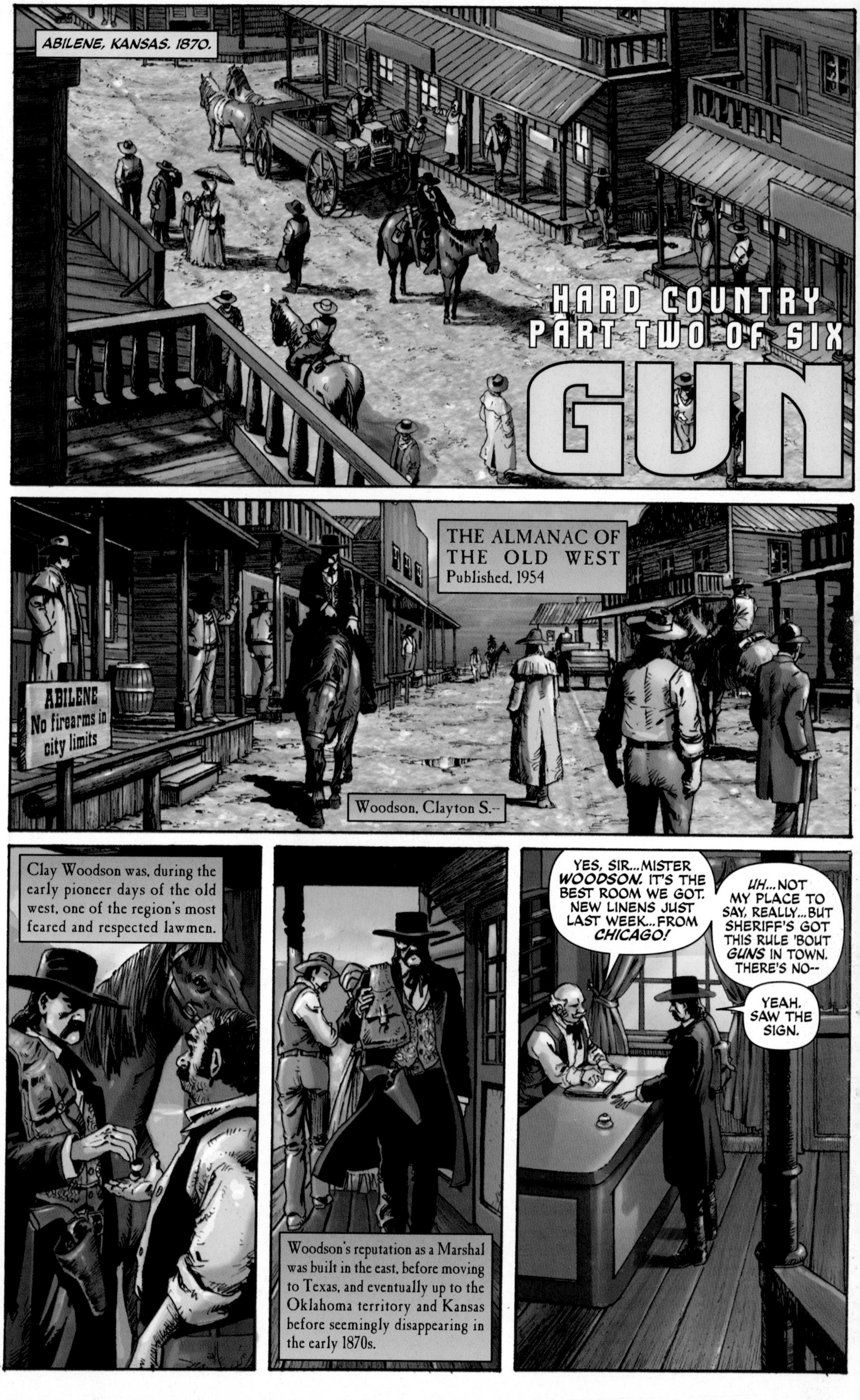
ABILENE, KANSAS. 1870.
HARD COUNTRY
PART TWO OF SIX
GUN
THE ALMANAC OF THE OLD WEST
Published, 1954
ABILENE
No firearms in city limits
Woodson, Clayton S.--
Clay Woodson was, during the early pioneer days of the old west, one of the region's most feared and respected lawmen.
Woodson's reputation as a Marshal was built in the east, before moving to Texas, and eventually up to the Oklahoma territory and Kansas before seemingly disappearing in the early 1870s.
YES, SIR...MISTER WOODSON. IT'S THE BEST ROOM WE GOT. NEW LINENS JUST LAST WEEK...FROM CHICAGO!
UH...NOT MY PLACE TO SAY, REALLY...BUT SHERIFF'S GOT THIS RULE 'BOUT GUNS IN TOWN. THERE'S NO--
YEAH. SAW THE SIGN.

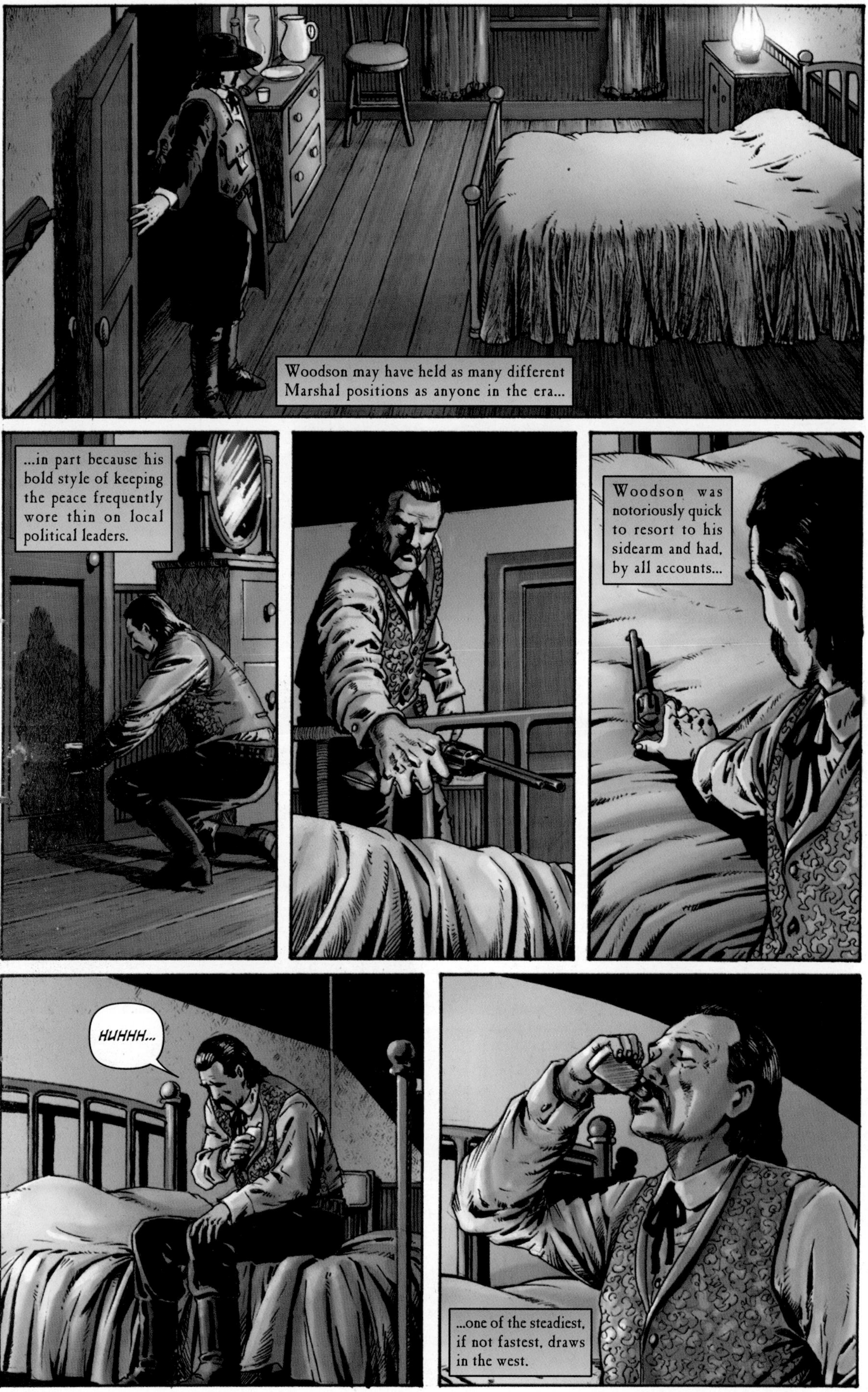
Woodson may have held as many different Marshal positions as anyone in the era...
...in part because his bold style of keeping the peace frequently wore thin on local political leaders.
Woodson was notoriously quick to resort to his sidearm and had, by all accounts...
HUHHH...
...one of the steadiest, if not fastest, draws in the west.

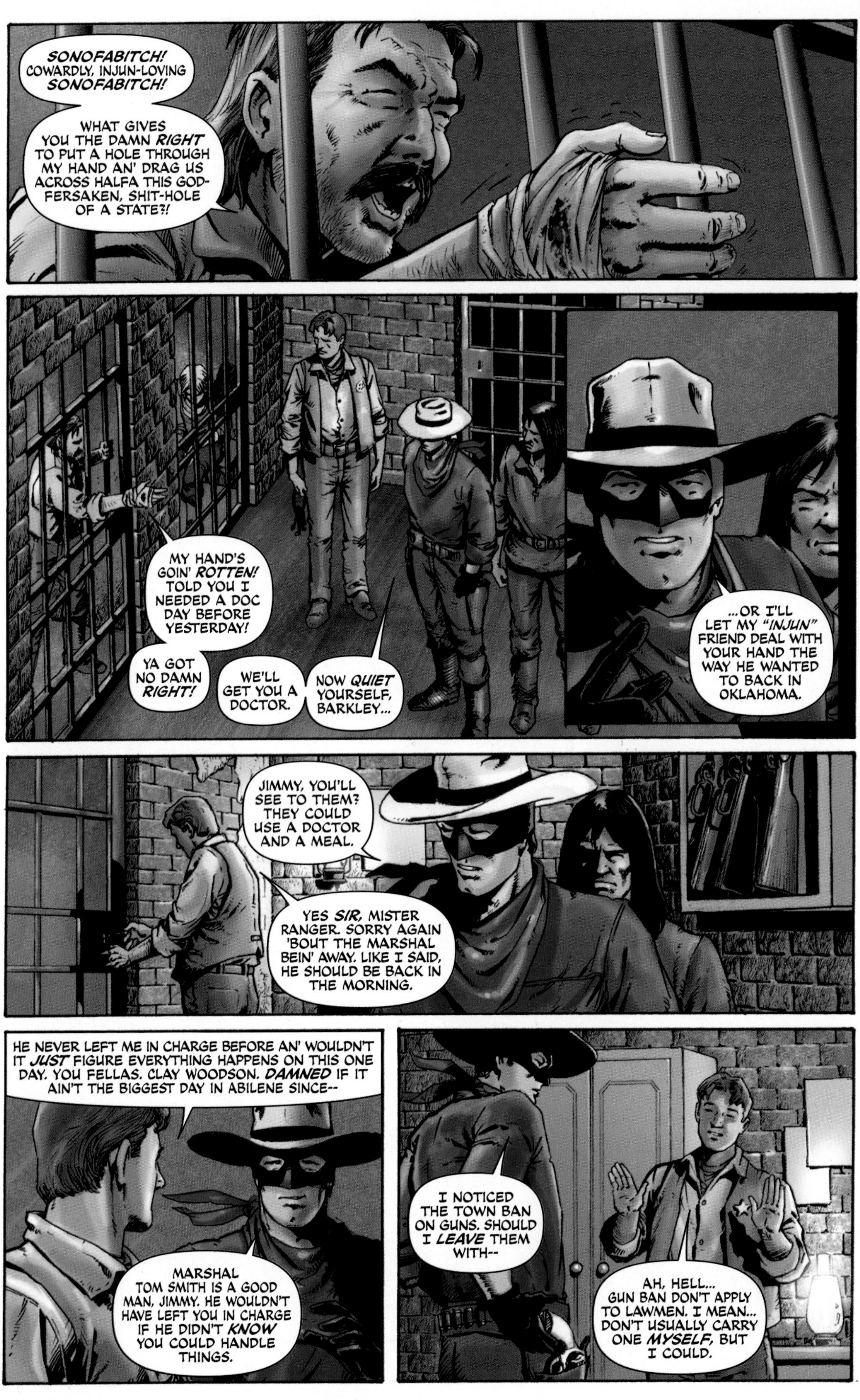
SONOFABITCH! COWARDLY, INJUN-LOVING SONOFABITCH!
WHAT GIVES YOU THE DAMN RIGHT TO PUT A HOLE THROUGH MY HAND AN' DRAG US ACROSS HALFA THIS GOD-FERSAKEN, SHIT-HOLE OF A STATE?!
MY HAND'S GOIN' ROTTEN! TOLD YOU I NEEDED A DOC DAY BEFORE YESTERDAY!
YA GOT NO DAMN RIGHT!
WE'LL GET YOU A DOCTOR.
NOW QUIET YOURSELF, BARKLEY...
...OR I'LL LET MY "INJUN" FRIEND DEAL WITH YOUR HAND THE WAY HE WANTED TO BACK IN OKLAHOMA.
JIMMY, YOU'LL SEE TO THEM? THEY COULD USE A DOCTOR AND A MEAL.
YES SIR, MISTER RANGER. SORRY AGAIN 'BOUT THE MARSHAL BEIN' AWAY. LIKE I SAID, HE SHOULD BE BACK IN THE MORNING.
HE NEVER LEFT ME IN CHARGE BEFORE AN' WOULDN'T IT JUST FIGURE EVERYTHING HAPPENS ON THIS ONE DAY. YOU FELLAS. CLAY WOODSON. DAMNED IF IT AIN'T THE BIGGEST DAY IN ABILENE SINCE--
MARSHAL TOM SMITH IS A GOOD MAN, JIMMY. HE WOULDN'T HAVE LEFT YOU IN CHARGE IF HE DIDN'T KNOW YOU COULD HANDLE THINGS.
I NOTICED THE TOWN BAN ON GUNS. SHOULD I LEAVE THEM WITH--
AH, HELL... GUN BAN DON'T APPLY TO LAWMEN. I MEAN... DON'T USUALLY CARRY ONE MYSELF, BUT I COULD.

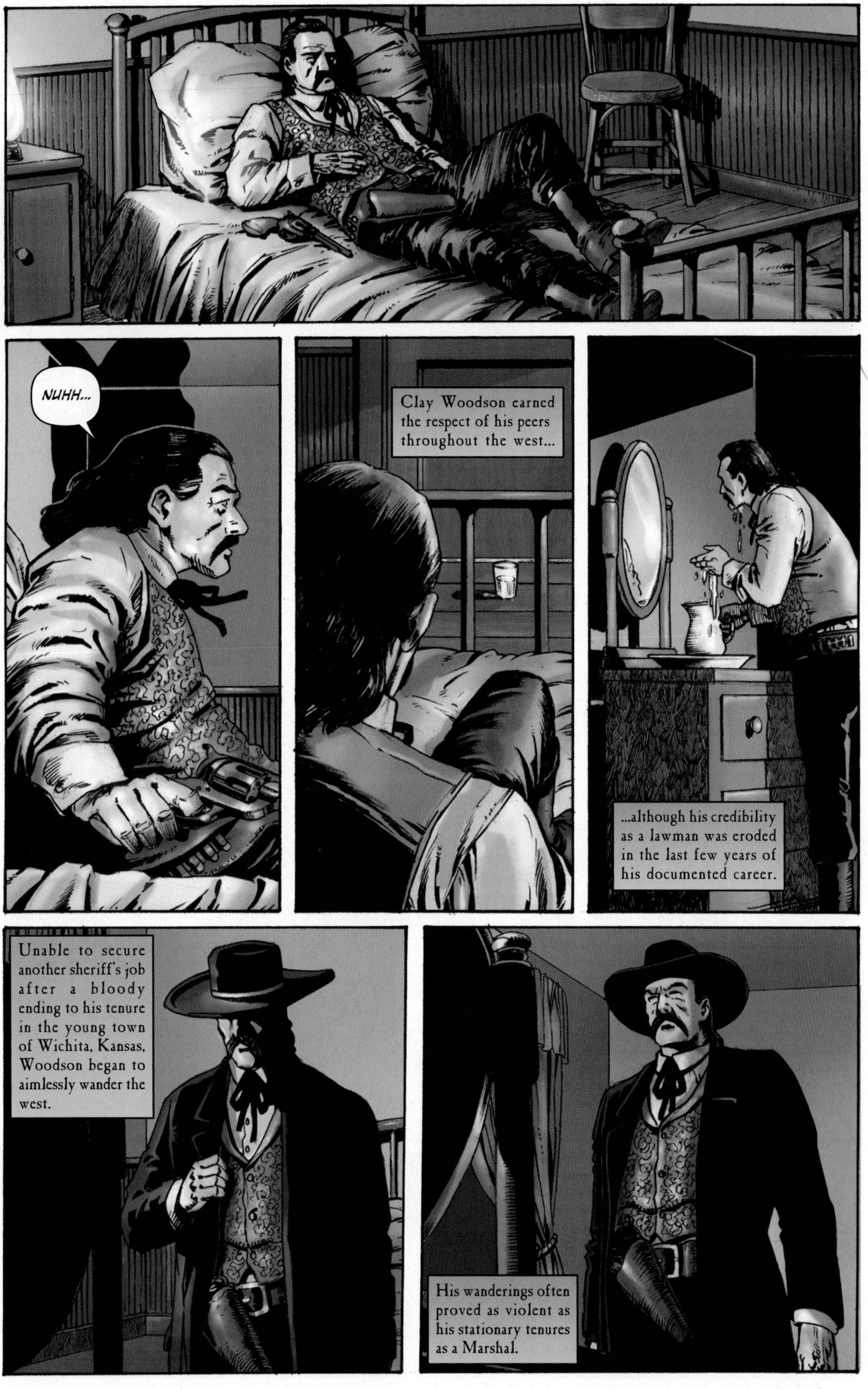

NUHH...
Clay Woodson earned the respect of his peers throughout the west...
...although his credibility as a lawman was eroded in the last few years of his documented career.
Unable to secure another sheriff's job after a bloody ending to his tenure in the young town of Wichita, Kansas, Woodson began to aimlessly wander the west.
His wanderings often proved as violent as his stationary tenures as a Marshal.

YOU GOT *REAL* IVORY GRIPS ON THEM GUNS?

YOU SHOOT BAD GUYS WITH THOSE THINGS?

WHERE'S *SILVER?*

PAPER SAYS YOU DON'T SHOOT *NOBODY,* SO WHAT'RE THEY FOR ANYWAY?

IT DON'T SAY HE DON'T SHOOT *NOBODY,* YOU IDJIT! SAYS HE DON'T *KILL* 'EM.

WHERE'S *SILVER?* IS HE REALLY TEN FEET TALL?

TEN? NO... I DON'T THINK SO. HE'D BE PRETTY HARD TO MOUNT.

SILVER'S RESTING. WE HAD A LONG, HARD TRIP FROM OKLAHOMA.

I SHOOT WHEN I HAVE TO. *ONLY* AT THOSE WHO DESERVE IT, AND *ONLY* TO DISARM. THE PAPER GOT THAT PART RIGHT, BOYS.

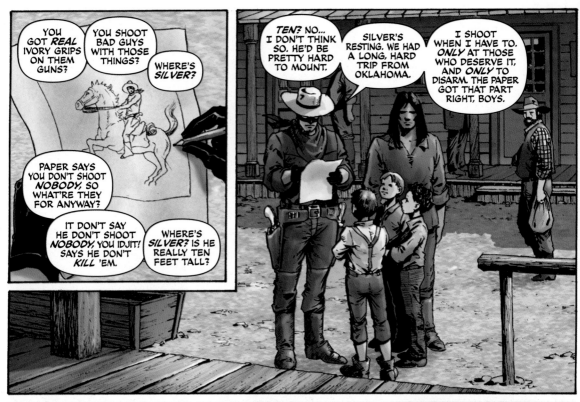

AND, *YES...* THEY'RE IVORY.

IVORY. *DAMN.* CAN I HOLD ONE?

WATCH YOUR MOUTH, SON. AND, NO... YOU CAN'T HOLD ONE.

ALL RIGHT, BOYS. TONTO AND I NEED TO MOVE ON.

C'MON... BART SAYS HIS PA SAW CLAY WOODSON AT THE CROWN.

HOT *DAMN...*WE GOT TIME BEFORE SUPPER!

CLAY WOODSON. MY FATHER USED TO TALK ABOUT HIM.

SAID HE WAS RUTHLESS, BUT A GOOD LAWMAN. *ONCE* WAS, ANYWAY...

MAYBE WE CAN GET HIS *AUTOGRAPH.*

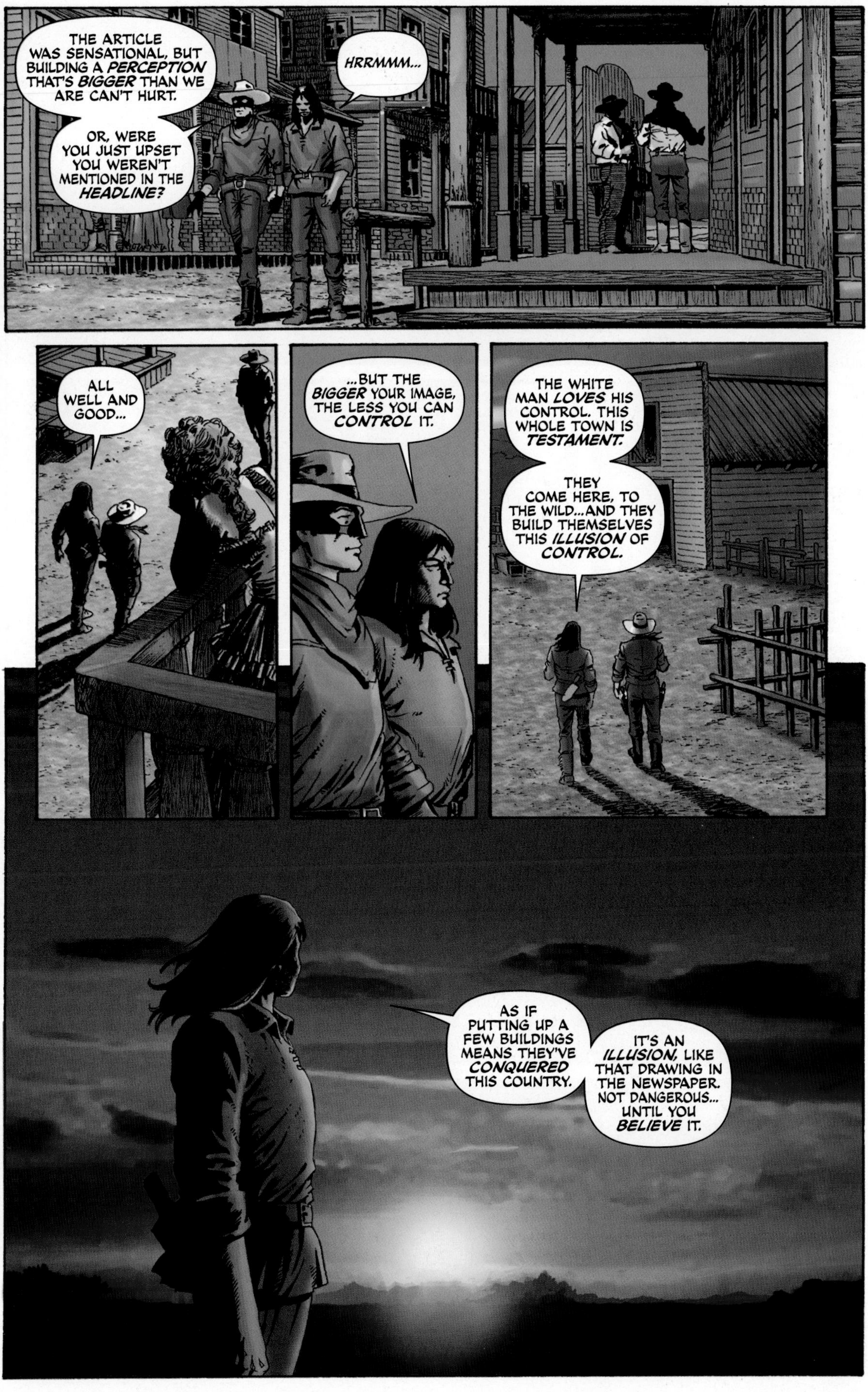
THE ARTICLE WAS SENSATIONAL, BUT BUILDING A PERCEPTION THAT'S BIGGER THAN WE ARE CAN'T HURT.
HRRMMM...
OR, WERE YOU JUST UPSET YOU WEREN'T MENTIONED IN THE HEADLINE?
ALL WELL AND GOOD...
...BUT THE BIGGER YOUR IMAGE, THE LESS YOU CAN CONTROL IT.
THE WHITE MAN LOVES HIS CONTROL. THIS WHOLE TOWN IS TESTAMENT.
THEY COME HERE, TO THE WILD...AND THEY BUILD THEMSELVES THIS ILLUSION OF CONTROL.
AS IF PUTTING UP A FEW BUILDINGS MEANS THEY'VE CONQUERED THIS COUNTRY.
IT'S AN ILLUSION, LIKE THAT DRAWING IN THE NEWSPAPER. NOT DANGEROUS... UNTIL YOU BELIEVE IT.

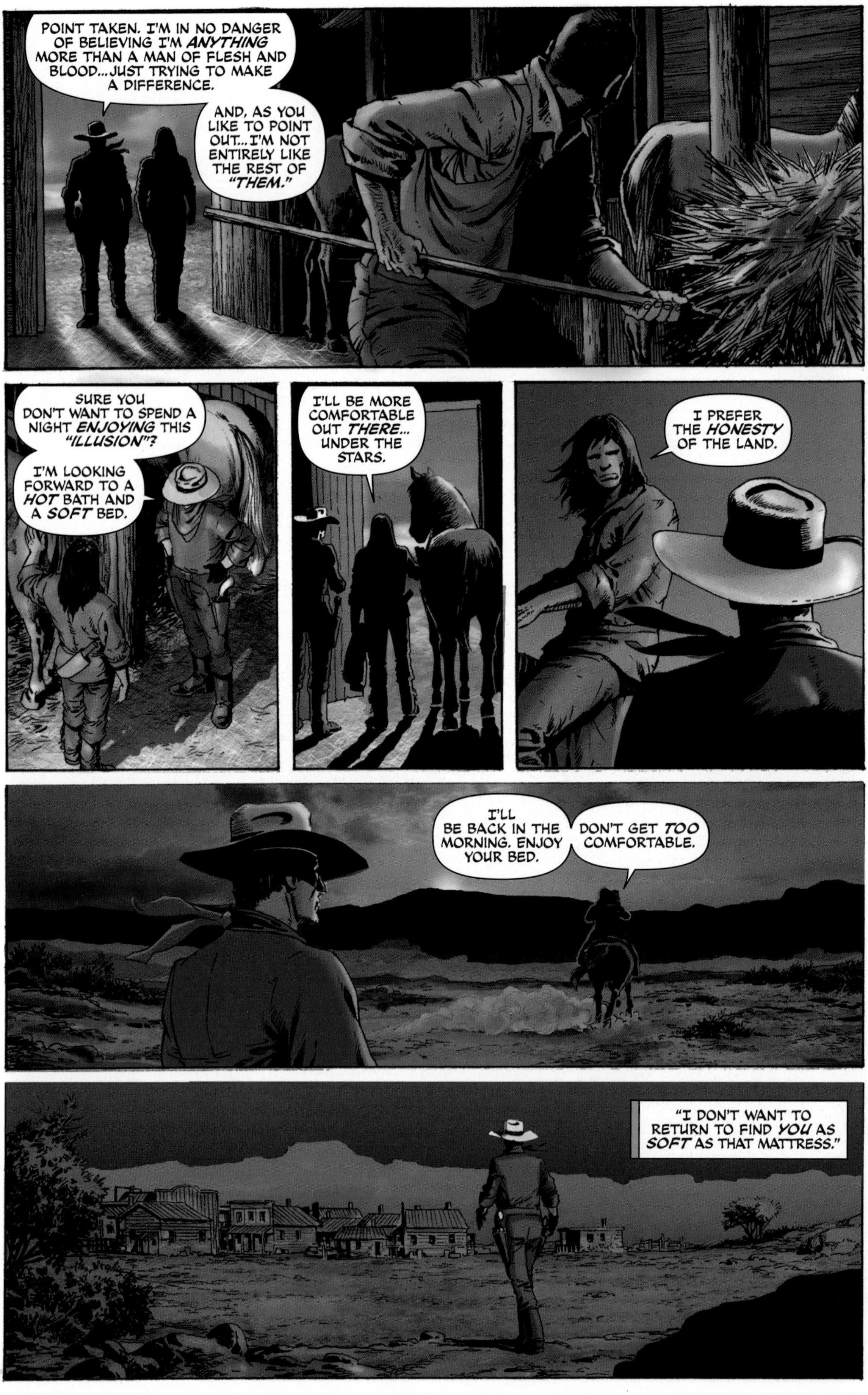
POINT TAKEN. I'M IN NO DANGER OF BELIEVING I'M ANYTHING MORE THAN A MAN OF FLESH AND BLOOD...JUST TRYING TO MAKE A DIFFERENCE.
AND, AS YOU LIKE TO POINT OUT...I'M NOT ENTIRELY LIKE THE REST OF "THEM."
SURE YOU DON'T WANT TO SPEND A NIGHT ENJOYING THIS "ILLUSION"?
I'M LOOKING FORWARD TO A HOT BATH AND A SOFT BED.
I'LL BE MORE COMFORTABLE OUT THERE... UNDER THE STARS.
I PREFER THE HONESTY OF THE LAND.
I'LL BE BACK IN THE MORNING. ENJOY YOUR BED.
DON'T GET TOO COMFORTABLE.
"I DON'T WANT TO RETURN TO FIND YOU AS SOFT AS THAT MATTRESS."

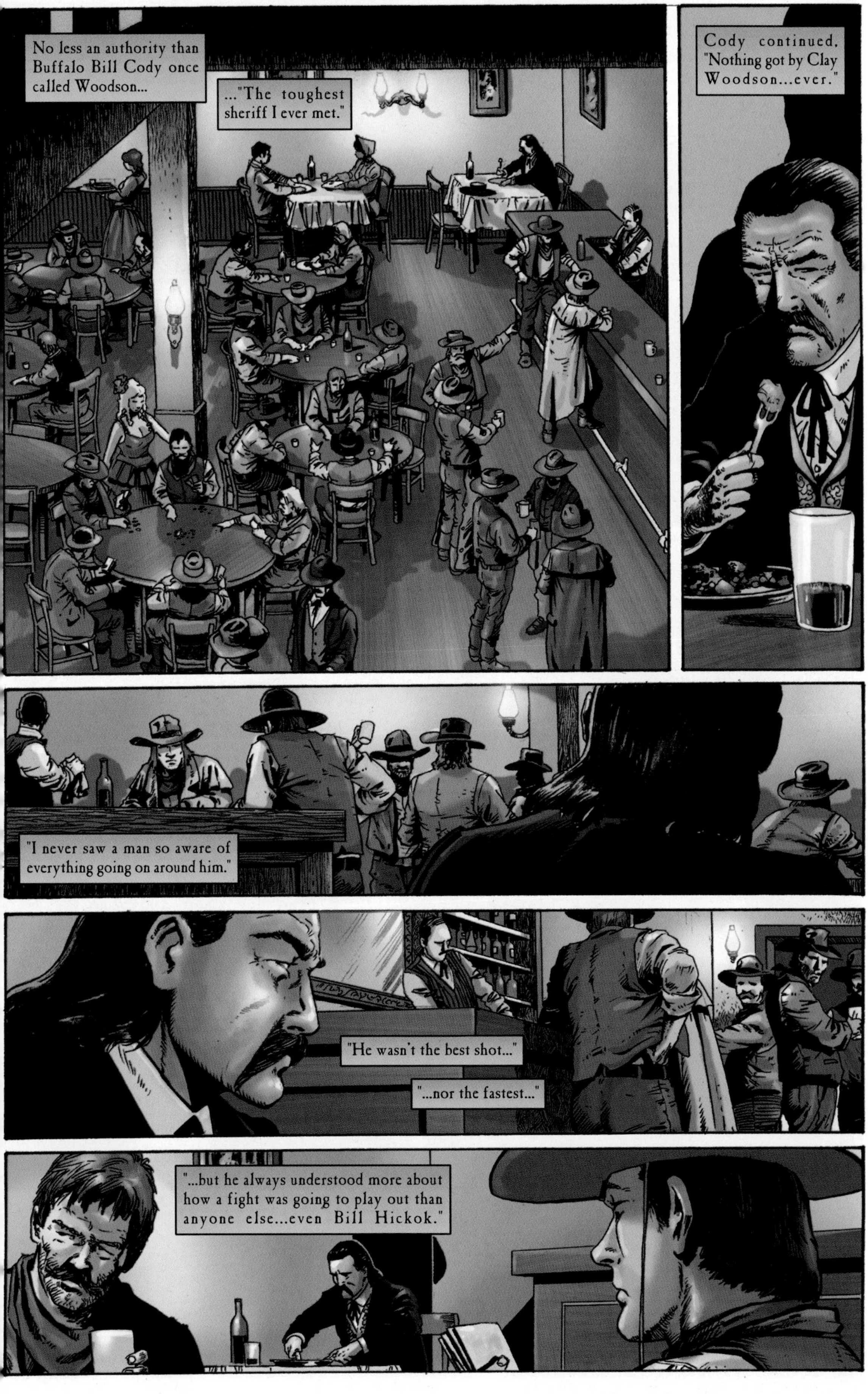

No less an authority than Buffalo Bill Cody once called Woodson...
..."The toughest sheriff I ever met."
Cody continued, "Nothing got by Clay Woodson...ever."
"I never saw a man so aware of everything going on around him."
"He wasn't the best shot..."
"...nor the fastest..."
"...but he always understood more about how a fight was going to play out than anyone else...even Bill Hickok."

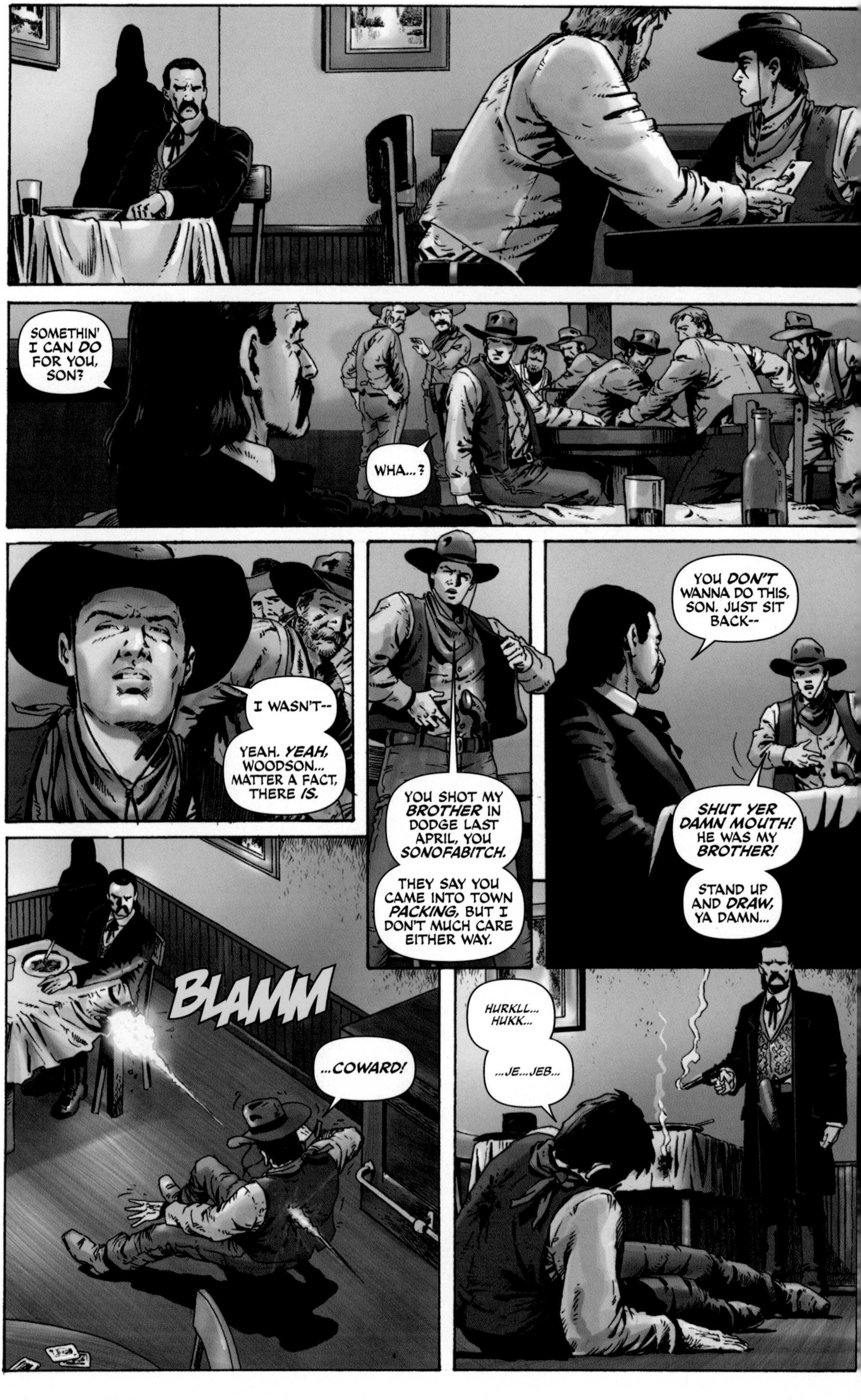
SOMETHIN' I CAN DO FOR YOU, SON?
WHA...?
I WASN'T--
YEAH. YEAH, WOODSON... MATTER A FACT, THERE IS.
YOU SHOT MY BROTHER IN DODGE LAST APRIL, YOU SONOFABITCH.
THEY SAY YOU CAME INTO TOWN PACKING, BUT I DON'T MUCH CARE EITHER WAY.
YOU DON'T WANNA DO THIS, SON. JUST SIT BACK--
SHUT YER DAMN MOUTH! HE WAS MY BROTHER!
STAND UP AND DRAW, YA DAMN...
BLAMM
...COWARD!
HURKLL... HUKK...
...JE...JEB...

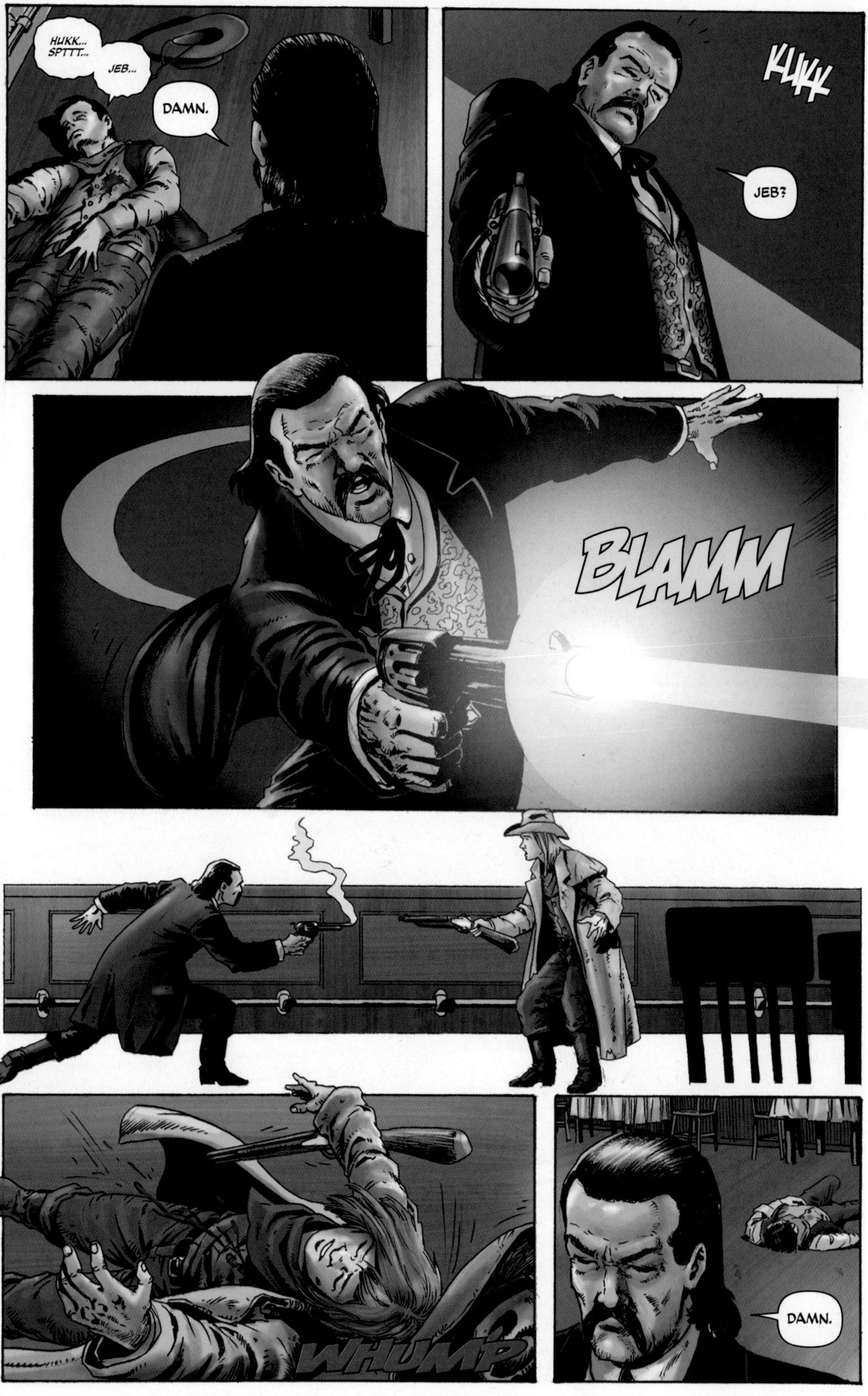
HUKK... SPTTT...
JEB...
DAMN.
KLIKK
JEB?
BLAMM
WHUMP
DAMN.

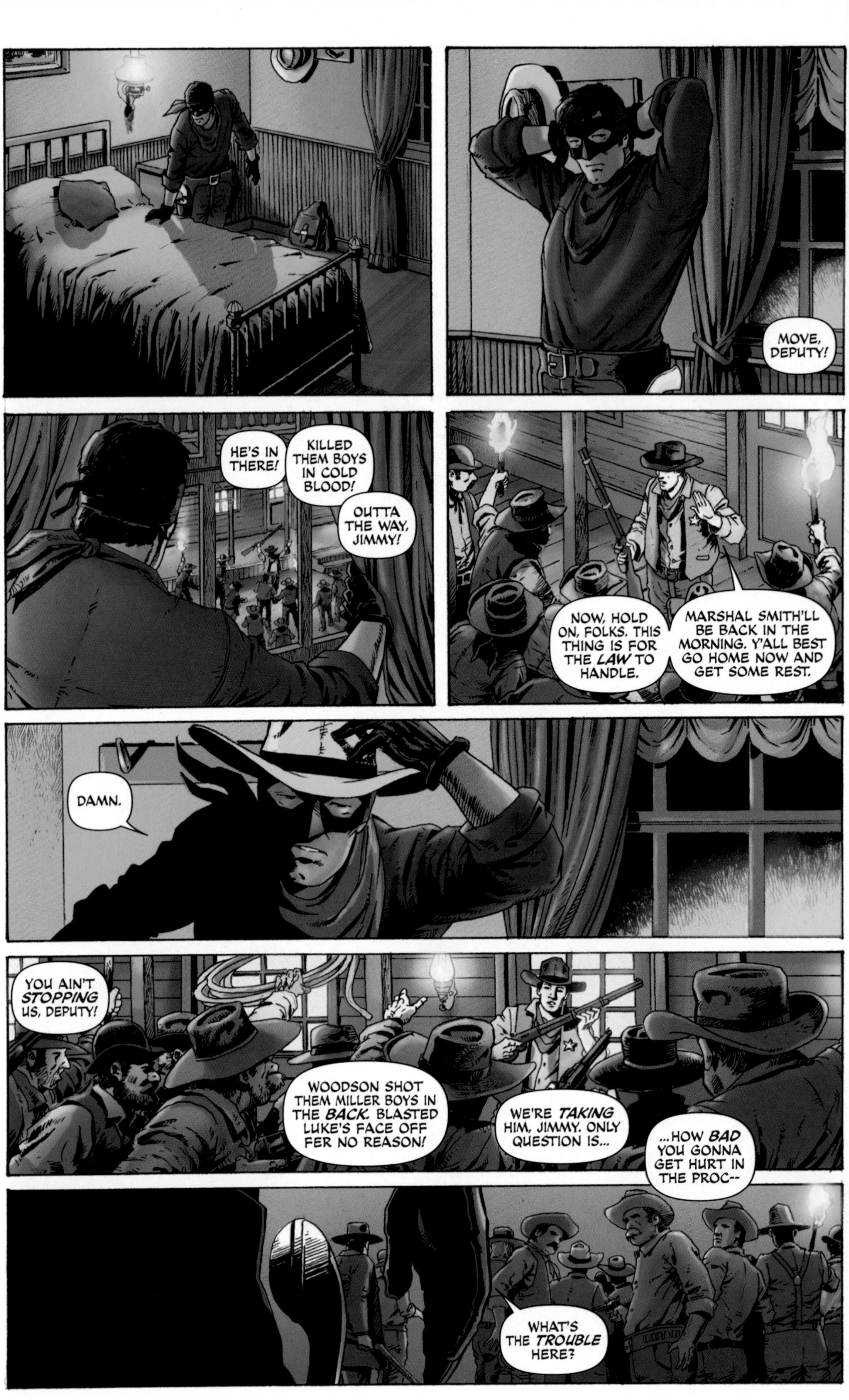
MOVE, DEPUTY!
HE'S IN THERE!
KILLED THEM BOYS IN COLD BLOOD!
OUTTA THE WAY, JIMMY!
NOW, HOLD ON, FOLKS. THIS THING IS FOR THE *LAW* TO HANDLE.
MARSHAL SMITH'LL BE BACK IN THE MORNING. Y'ALL BEST GO HOME NOW AND GET SOME REST.
DAMN.
YOU AIN'T *STOPPING* US, DEPUTY!
WOODSON SHOT THEM MILLER BOYS IN THE *BACK*. BLASTED LUKE'S FACE OFF FER NO REASON!
WE'RE *TAKING* HIM, JIMMY. ONLY QUESTION IS...
...HOW *BAD* YOU GONNA GET HURT IN THE PROC--
WHAT'S THE *TROUBLE* HERE?

I BELIEVE THE DEPUTY WAS TRYING TO BE HEARD.

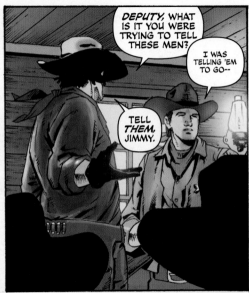

DEPUTY, WHAT IS IT YOU WERE TRYING TO TELL THESE MEN?

I WAS TELLING 'EM TO GO--

TELL *THEM*, JIMMY.

LOOK, I KNOW SOME OF YOU MIGHT HAVE *KIN* IN THERE. MAYBE THEY BEEN *HURT* OR *KILLED*, BUT THIS IS A MATTER FOR THE *LAW* NOW.

JUS' GIVE US SOME TIME TO SEE WHAT'S HAPPENED AND SORT THIS THING OUT.

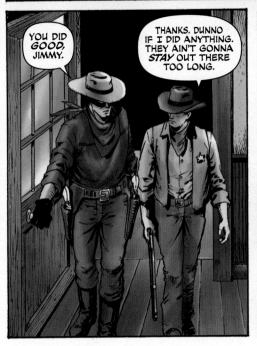

YOU DID *GOOD*, JIMMY.

THANKS. DUNNO IF I DID ANYTHING. THEY AIN'T GONNA *STAY* OUT THERE TOO LONG.

WE'D BEST *FIND* WOODSON 'FORE--

GENTLEMEN.

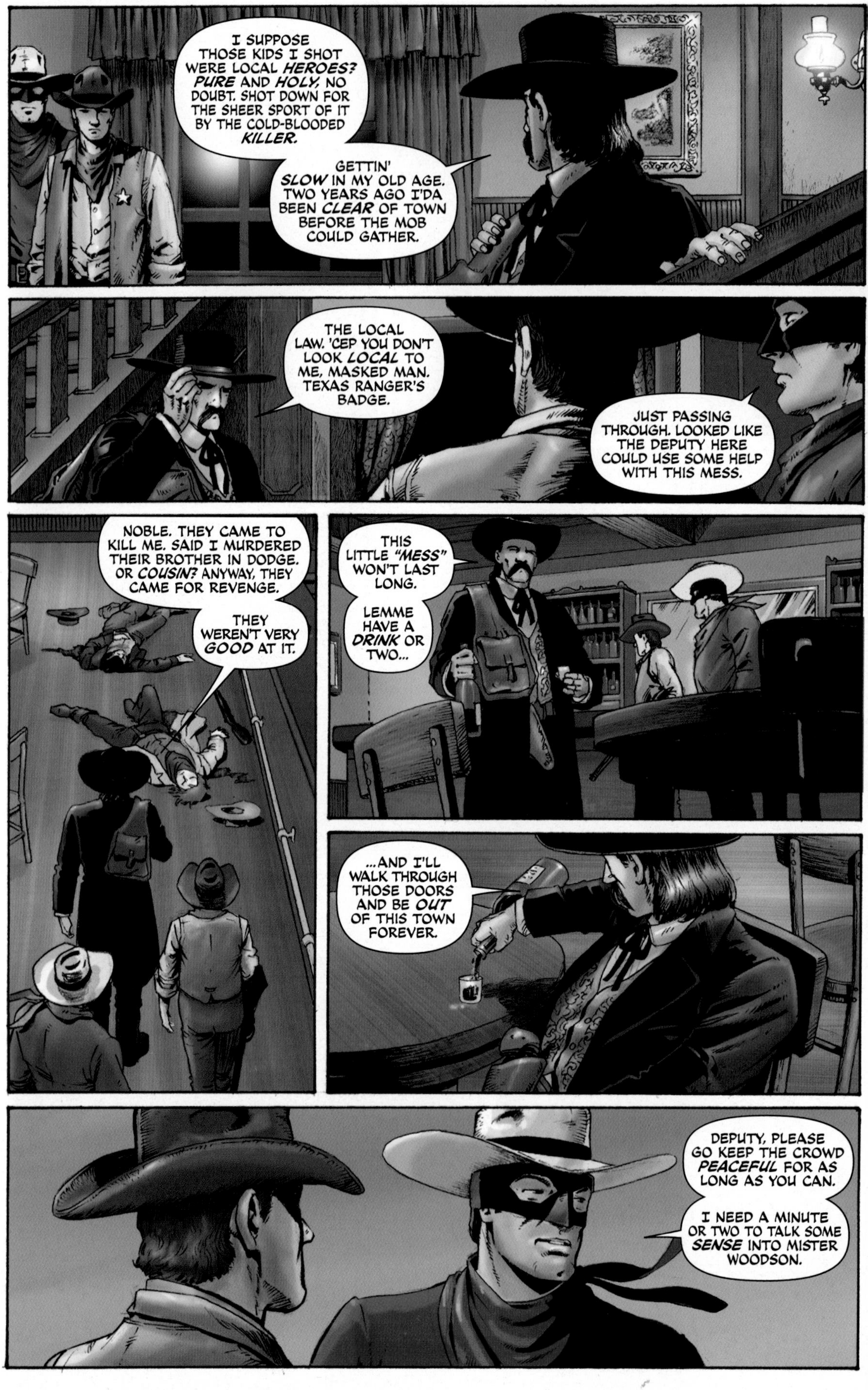
I SUPPOSE THOSE KIDS I SHOT WERE LOCAL HEROES? PURE AND HOLY, NO DOUBT. SHOT DOWN FOR THE SHEER SPORT OF IT BY THE COLD-BLOODED KILLER.
GETTIN' SLOW IN MY OLD AGE. TWO YEARS AGO I'DA BEEN CLEAR OF TOWN BEFORE THE MOB COULD GATHER.
THE LOCAL LAW. 'CEP YOU DON'T LOOK LOCAL TO ME, MASKED MAN. TEXAS RANGER'S BADGE.
JUST PASSING THROUGH. LOOKED LIKE THE DEPUTY HERE COULD USE SOME HELP WITH THIS MESS.
NOBLE. THEY CAME TO KILL ME. SAID I MURDERED THEIR BROTHER IN DODGE. OR COUSIN? ANYWAY, THEY CAME FOR REVENGE.
THEY WEREN'T VERY GOOD AT IT.
THIS LITTLE "MESS" WON'T LAST LONG.
LEMME HAVE A DRINK OR TWO...
...AND I'LL WALK THROUGH THOSE DOORS AND BE OUT OF THIS TOWN FOREVER.
DEPUTY, PLEASE GO KEEP THE CROWD PEACEFUL FOR AS LONG AS YOU CAN.
I NEED A MINUTE OR TWO TO TALK SOME SENSE INTO MISTER WOODSON.

YOU SPENT A LOT OF YEARS AS A MARSHAL.

YOU *KNOW* YOU CAN'T WALK FROM THIS WITHOUT AT *LEAST* A HEARING.

IF IT HAPPENED LIKE YOU SAY...IF *THEY* CAME AFTER *YOU*, WHICH IT APPEARS THEY DID...YOU DON'T HAVE ANYTHING TO WORRY ABOUT.

HAHA...NOTHING AT ALL, UNTIL THE *JUDGE* TURNS OUT TO BE THAT KID'S *UNCLE.*

AND THE *JURY* IS FULL OF *BROTHERS* AND *COUSINS*...ALL *DYING* TO SEE MY NECK A FOOT LONGER.

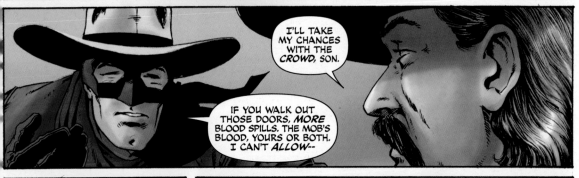

I'LL TAKE MY CHANCES WITH THE *CROWD,* SON.

IF YOU WALK OUT THOSE DOORS, *MORE* BLOOD SPILLS. THE MOB'S BLOOD, YOURS OR BOTH. I CAN'T *ALLOW--*

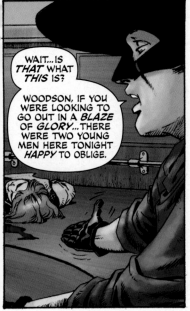

WAIT...IS *THAT* WHAT *THIS* IS?

WOODSON, IF YOU WERE LOOKING TO GO OUT IN A *BLAZE* OF *GLORY*...THERE WERE TWO YOUNG MEN HERE TONIGHT *HAPPY* TO OBLIGE.

EVER *KILLED* A MAN, RANGER?

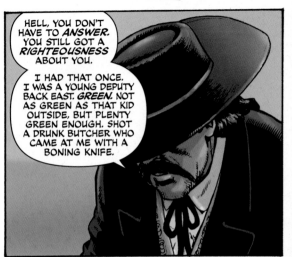

HELL, YOU DON'T HAVE TO *ANSWER.* YOU STILL GOT A *RIGHTEOUSNESS* ABOUT YOU.

I HAD THAT ONCE. I WAS A YOUNG DEPUTY BACK EAST. *GREEN.* NOT AS GREEN AS THAT KID OUTSIDE, BUT PLENTY GREEN ENOUGH. SHOT A DRUNK BUTCHER WHO CAME AT ME WITH A BONING KNIFE.

I *STILL* FELT RIGHTEOUS AFTER THE FIRST ONE. HELL, I STILL FELT *RIGHTEOUS* AFTER THE FIRST *DOZEN.*

SOMEWHERE AFTER THAT, I STOPPED THINKING MUCH ABOUT WHETHER THEY *DESERVED* IT.

I LOOKED AROUND ONE DAY AND I WASN'T A *MARSHAL* ANYMORE. I WASN'T A *HUSBAND* OR A *FATHER* OR EVEN A *SON.*

I WAS JUST THE OLD *SONOFABITCH* WHO WAS REALLY GOOD AT *KILLIN'.*

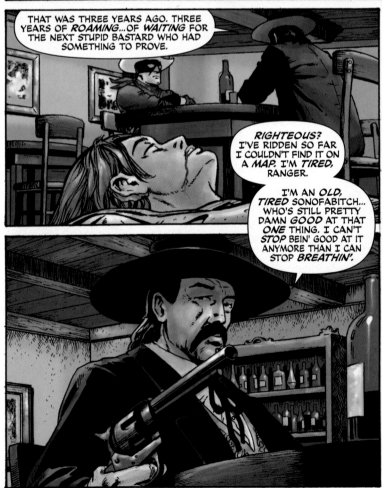

THAT WAS THREE YEARS AGO. THREE YEARS OF *ROAMING...* OF *WAITING* FOR THE NEXT STUPID BASTARD WHO HAD SOMETHING TO PROVE.

RIGHTEOUS? I'VE RIDDEN SO FAR I COULDN'T FIND IT ON A *MAP.* I'M *TIRED,* RANGER.

I'M AN *OLD, TIRED* SONOFABITCH... WHO'S STILL PRETTY DAMN *GOOD* AT THAT *ONE* THING. I CAN'T *STOP* BEIN' GOOD AT IT ANYMORE THAN I CAN STOP *BREATHIN'.*

I *SHOT* THEM BOYS FOR THE *SAME* REASON I'M WALKING OUT THERE NOW...

...'CAUSE IT'S WHAT I *DO.*

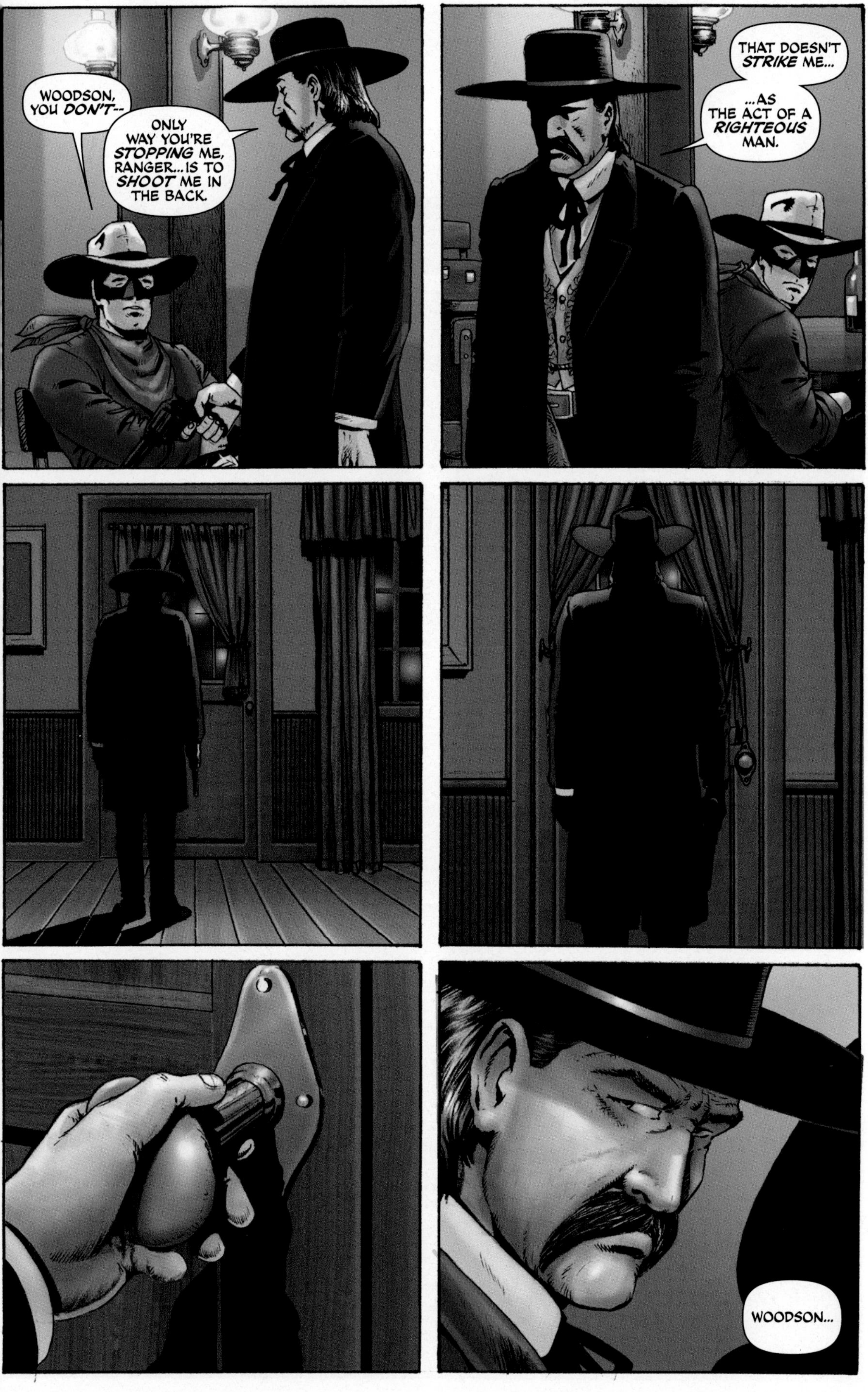
WOODSON, YOU DON'T--
ONLY WAY YOU'RE STOPPING ME, RANGER...IS TO SHOOT ME IN THE BACK.
THAT DOESN'T STRIKE ME...
...AS THE ACT OF A RIGHTEOUS MAN.
WOODSON...

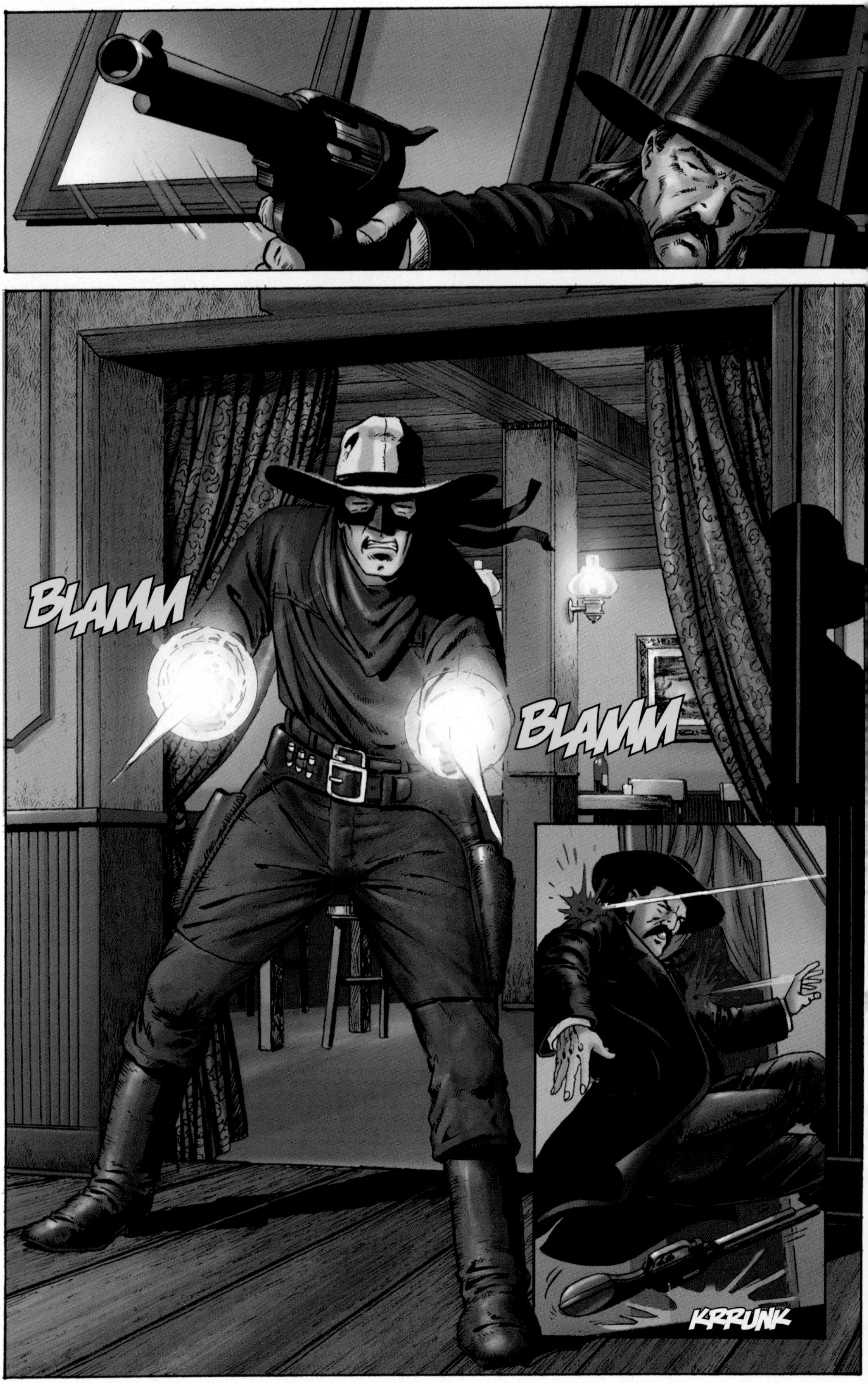
BLAMM
BLAMM
KRRUNK

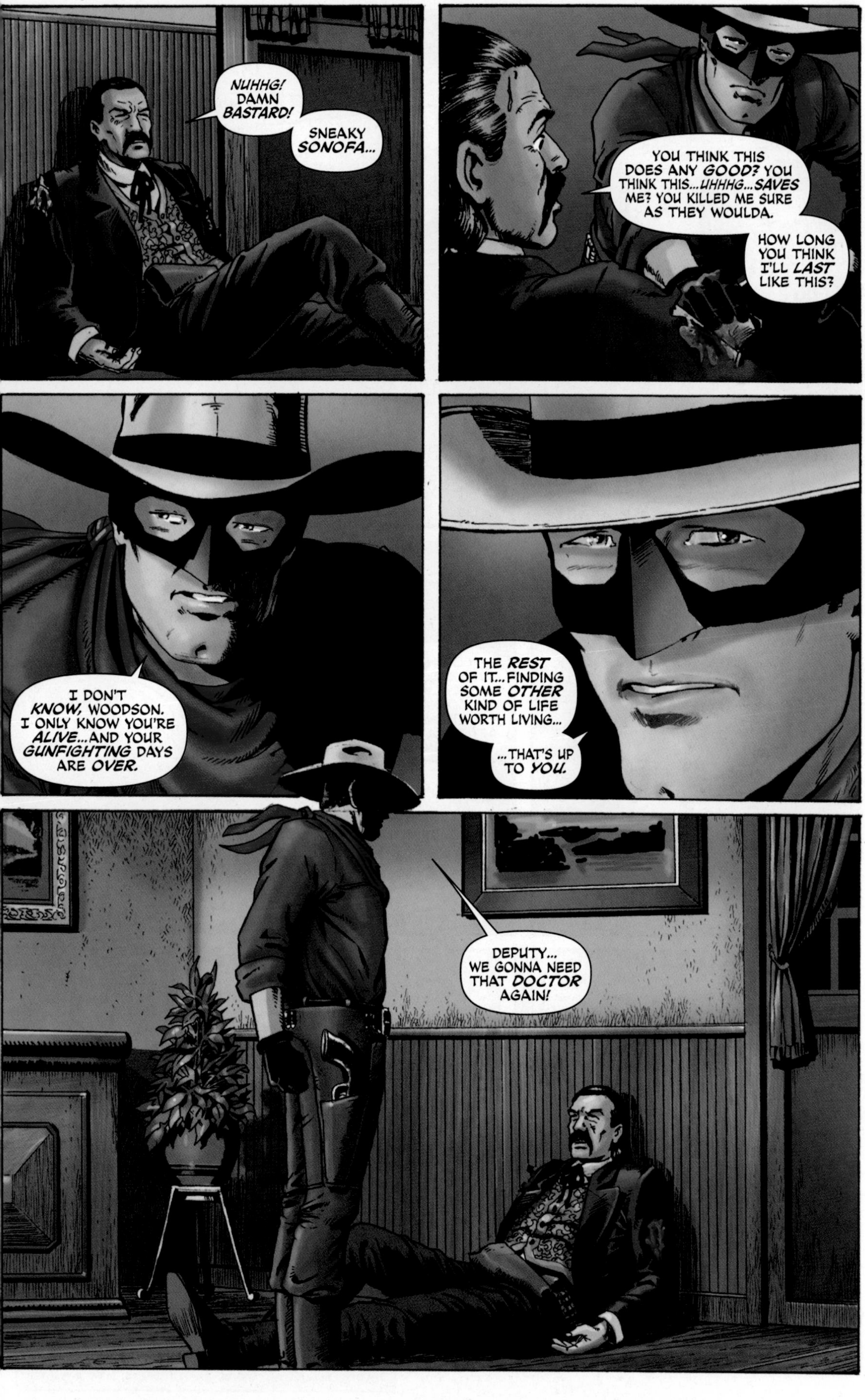
NUHHG! DAMN BASTARD!
SNEAKY SONOFA...
YOU THINK THIS DOES ANY GOOD? YOU THINK THIS...UHHHG...SAVES ME? YOU KILLED ME SURE AS THEY WOULDA.
HOW LONG YOU THINK I'LL LAST LIKE THIS?
I DON'T KNOW, WOODSON. I ONLY KNOW YOU'RE ALIVE...AND YOUR GUNFIGHTING DAYS ARE OVER.
THE REST OF IT...FINDING SOME OTHER KIND OF LIFE WORTH LIVING...
...THAT'S UP TO YOU.
DEPUTY... WE GONNA NEED THAT DOCTOR AGAIN!

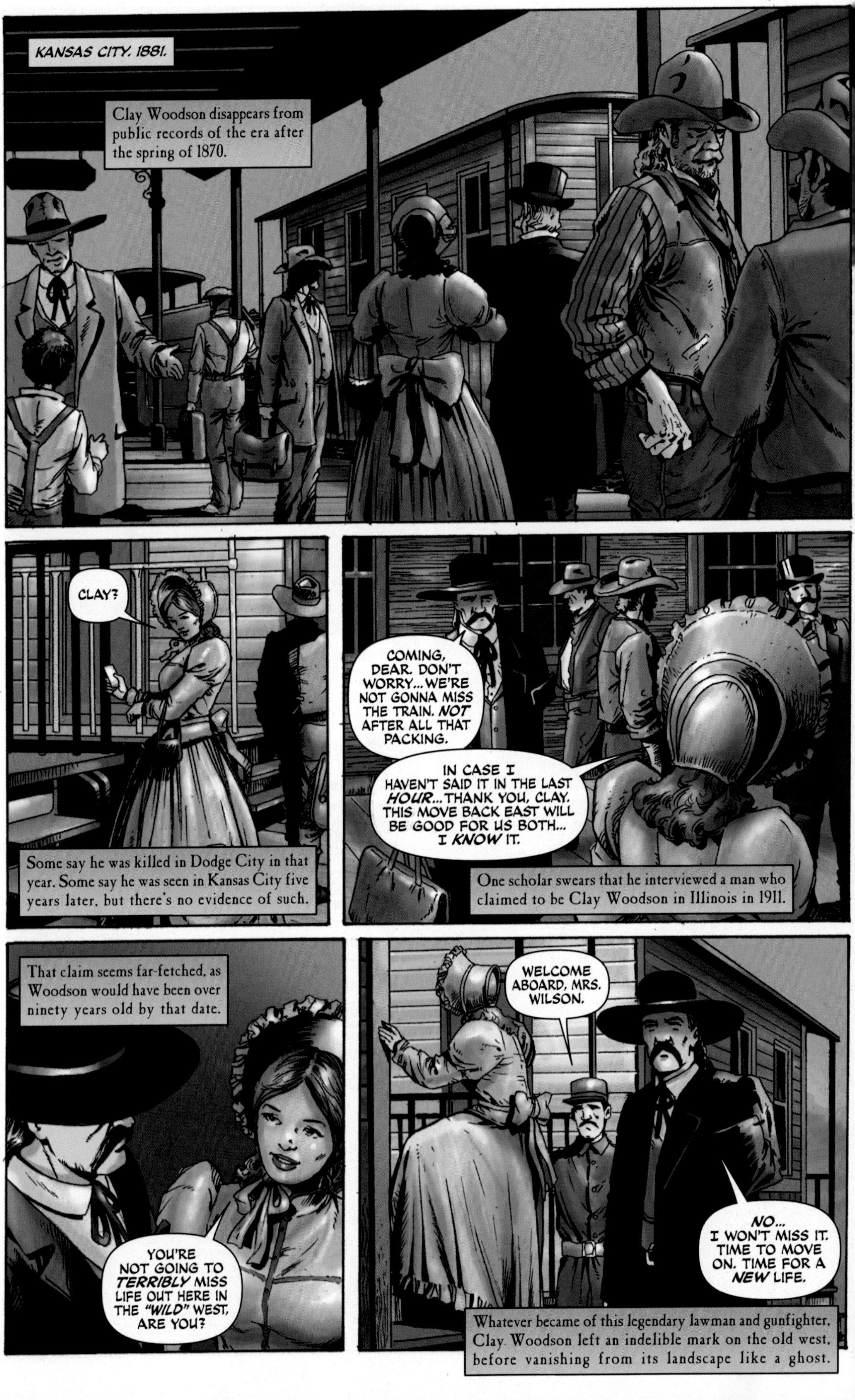
KANSAS CITY. 1881.
Clay Woodson disappears from public records of the era after the spring of 1870.
CLAY?
Some say he was killed in Dodge City in that year. Some say he was seen in Kansas City five years later, but there's no evidence of such.
COMING, DEAR. DON'T WORRY... WE'RE NOT GONNA MISS THE TRAIN. NOT AFTER ALL THAT PACKING.
IN CASE I HAVEN'T SAID IT IN THE LAST HOUR... THANK YOU, CLAY. THIS MOVE BACK EAST WILL BE GOOD FOR US BOTH... I KNOW IT.
One scholar swears that he interviewed a man who claimed to be Clay Woodson in Illinois in 1911.
That claim seems far-fetched, as Woodson would have been over ninety years old by that date.
YOU'RE NOT GOING TO TERRIBLY MISS LIFE OUT HERE IN THE "WILD" WEST, ARE YOU?
WELCOME ABOARD, MRS. WILSON.
NO... I WON'T MISS IT. TIME TO MOVE ON. TIME FOR A NEW LIFE.
Whatever became of this legendary lawman and gunfighter, Clay Woodson left an indelible mark on the old west, before vanishing from its landscape like a ghost.

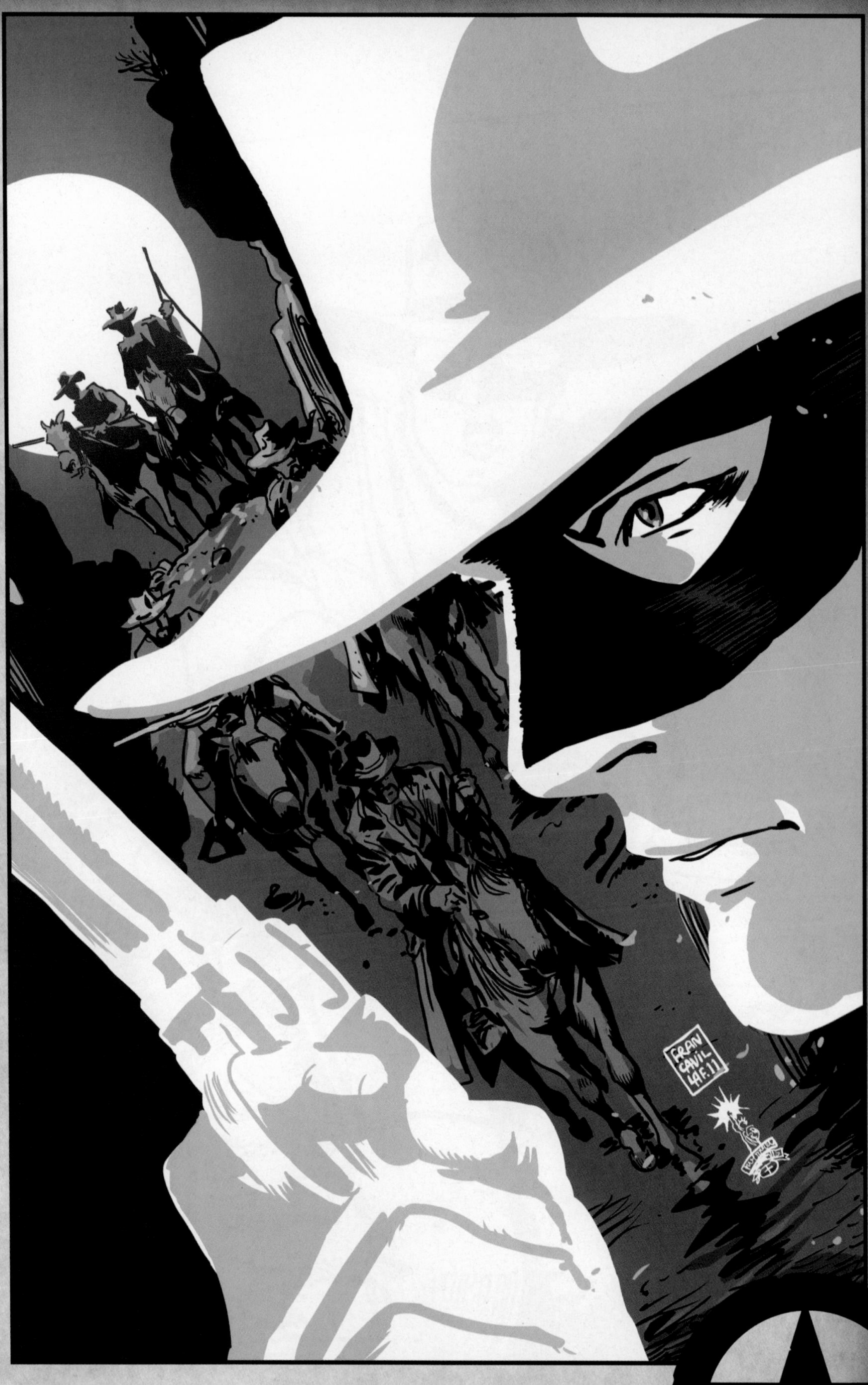

COVER BY **FRANCESCO FRANCAVILLA**

3

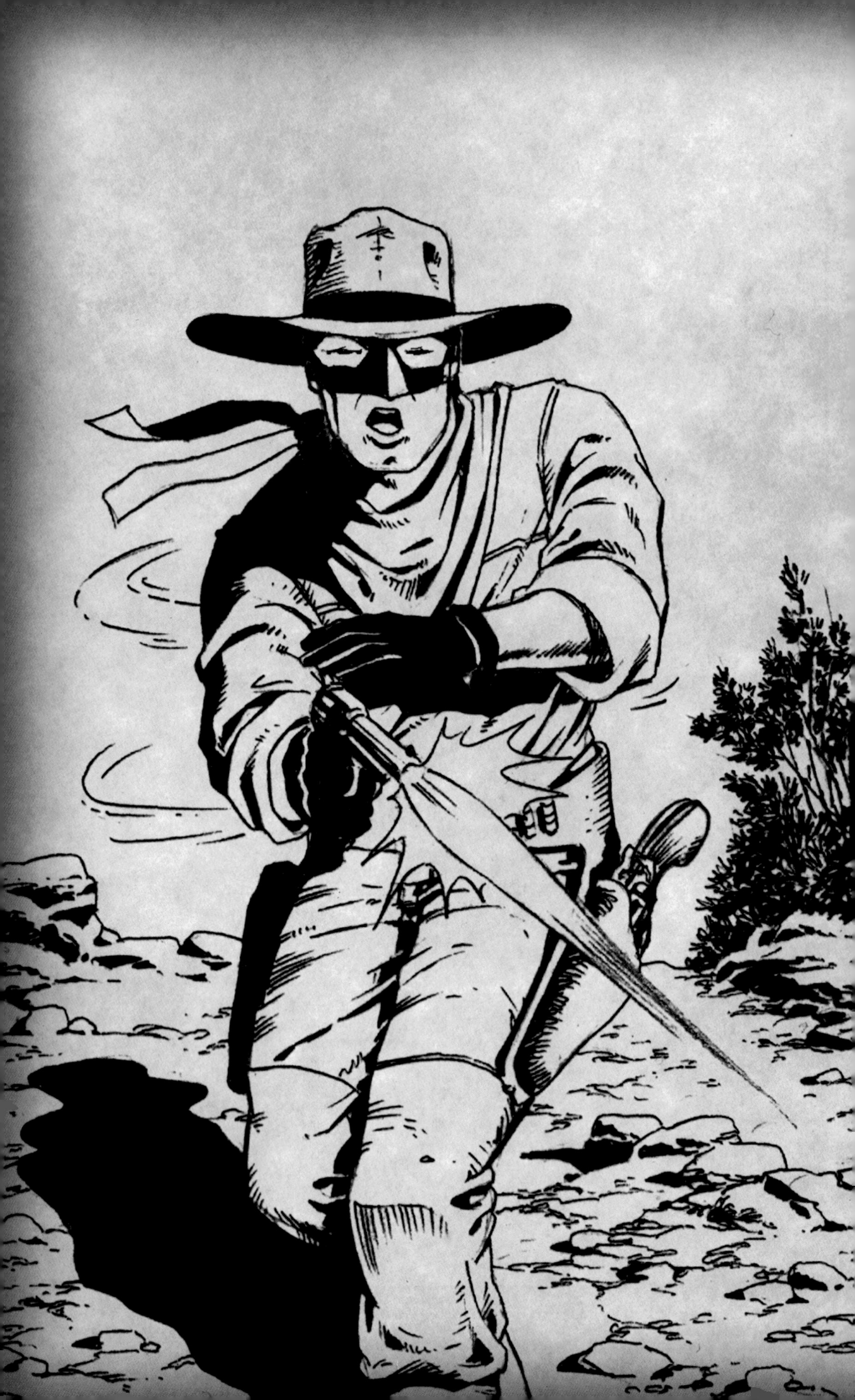

NEAR THE SOUTHERN EDGE OF THE COLORADO TERRITORY. 1870.

To trust the God of the Bible is to trust an irascible, vindictive, fierce and ever fickle and changeful master.

To trust the true God is to trust a Being who has uttered no promises.

UTOPIA
GOOD FOLK, FOREVER MINDFUL OF GODS WRATH

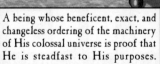
A being whose beneficent, exact, and changeless ordering of the machinery of His colossal universe is proof that He is steadfast to His purposes.

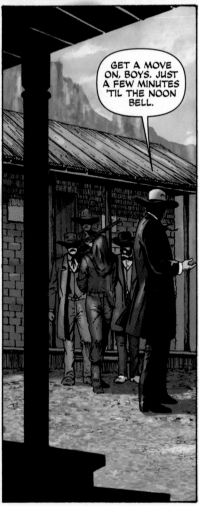

GET A MOVE ON, BOYS. JUST A FEW MINUTES 'TIL THE NOON BELL.

I WANT HIM *SWINGING* BY THE TIME THAT TWELFTH BELL SOUNDS.

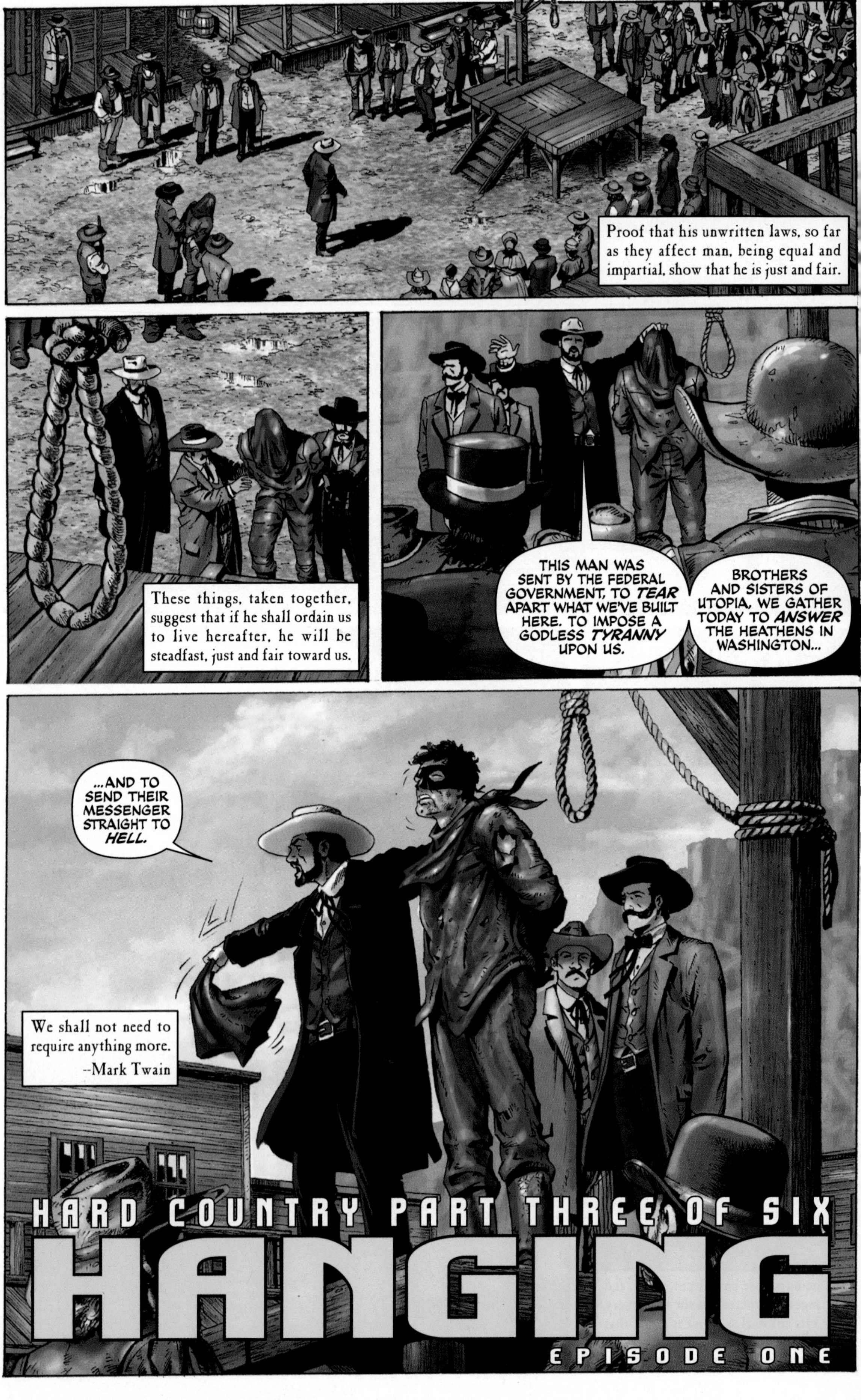
Proof that his unwritten laws, so far as they affect man, being equal and impartial, show that he is just and fair.
These things, taken together, suggest that if he shall ordain us to live hereafter, he will be steadfast, just and fair toward us.
BROTHERS AND SISTERS OF UTOPIA, WE GATHER TODAY TO ANSWER THE HEATHENS IN WASHINGTON...
THIS MAN WAS SENT BY THE FEDERAL GOVERNMENT, TO TEAR APART WHAT WE'VE BUILT HERE. TO IMPOSE A GODLESS TYRANNY UPON US.
...AND TO SEND THEIR MESSENGER STRAIGHT TO HELL.
We shall not need to require anything more. --Mark Twain
HARD COUNTRY PART THREE OF SIX
HANGING
EPISODE ONE

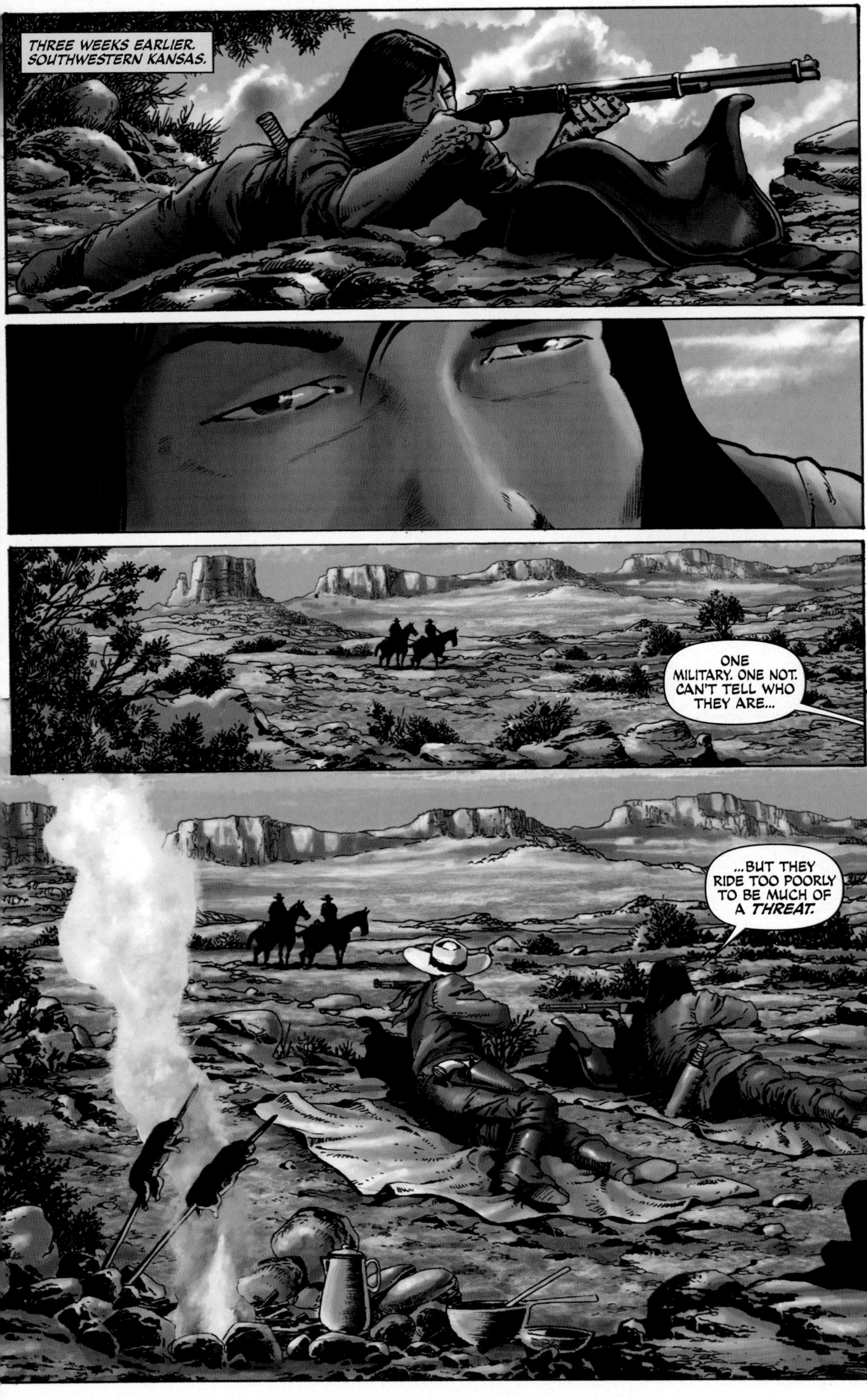
THREE WEEKS EARLIER. SOUTHWESTERN KANSAS.
ONE MILITARY. ONE NOT. CAN'T TELL WHO THEY ARE...
...BUT THEY RIDE TOO POORLY TO BE MUCH OF A *THREAT.*

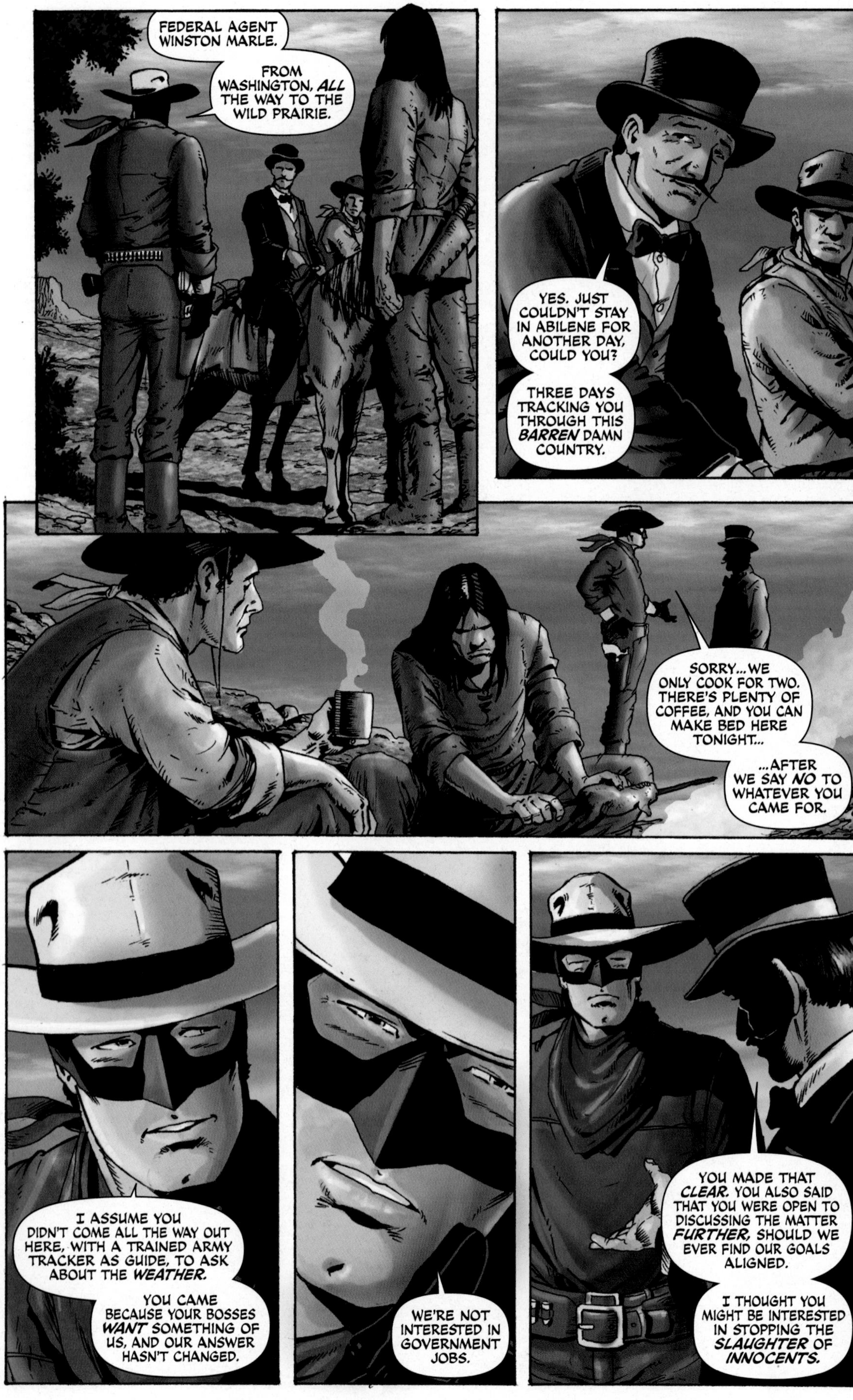
FEDERAL AGENT WINSTON MARLE.
FROM WASHINGTON, *ALL* THE WAY TO THE WILD PRAIRIE.
YES. JUST COULDN'T STAY IN ABILENE FOR ANOTHER DAY, COULD YOU?
THREE DAYS TRACKING YOU THROUGH THIS *BARREN* DAMN COUNTRY.
SORRY...WE ONLY COOK FOR TWO. THERE'S PLENTY OF COFFEE, AND YOU CAN MAKE BED HERE TONIGHT...
...AFTER WE SAY *NO* TO WHATEVER YOU CAME FOR.
I ASSUME YOU DIDN'T COME ALL THE WAY OUT HERE, WITH A TRAINED ARMY TRACKER AS GUIDE, TO ASK ABOUT THE *WEATHER.*
YOU CAME BECAUSE YOUR BOSSES *WANT* SOMETHING OF US, AND OUR ANSWER HASN'T CHANGED.
WE'RE NOT INTERESTED IN GOVERNMENT JOBS.
YOU MADE THAT *CLEAR.* YOU ALSO SAID THAT YOU WERE OPEN TO DISCUSSING THE MATTER *FURTHER,* SHOULD WE EVER FIND OUR GOALS ALIGNED.
I THOUGHT YOU MIGHT BE INTERESTED IN STOPPING THE *SLAUGHTER* OF *INNOCENTS.*

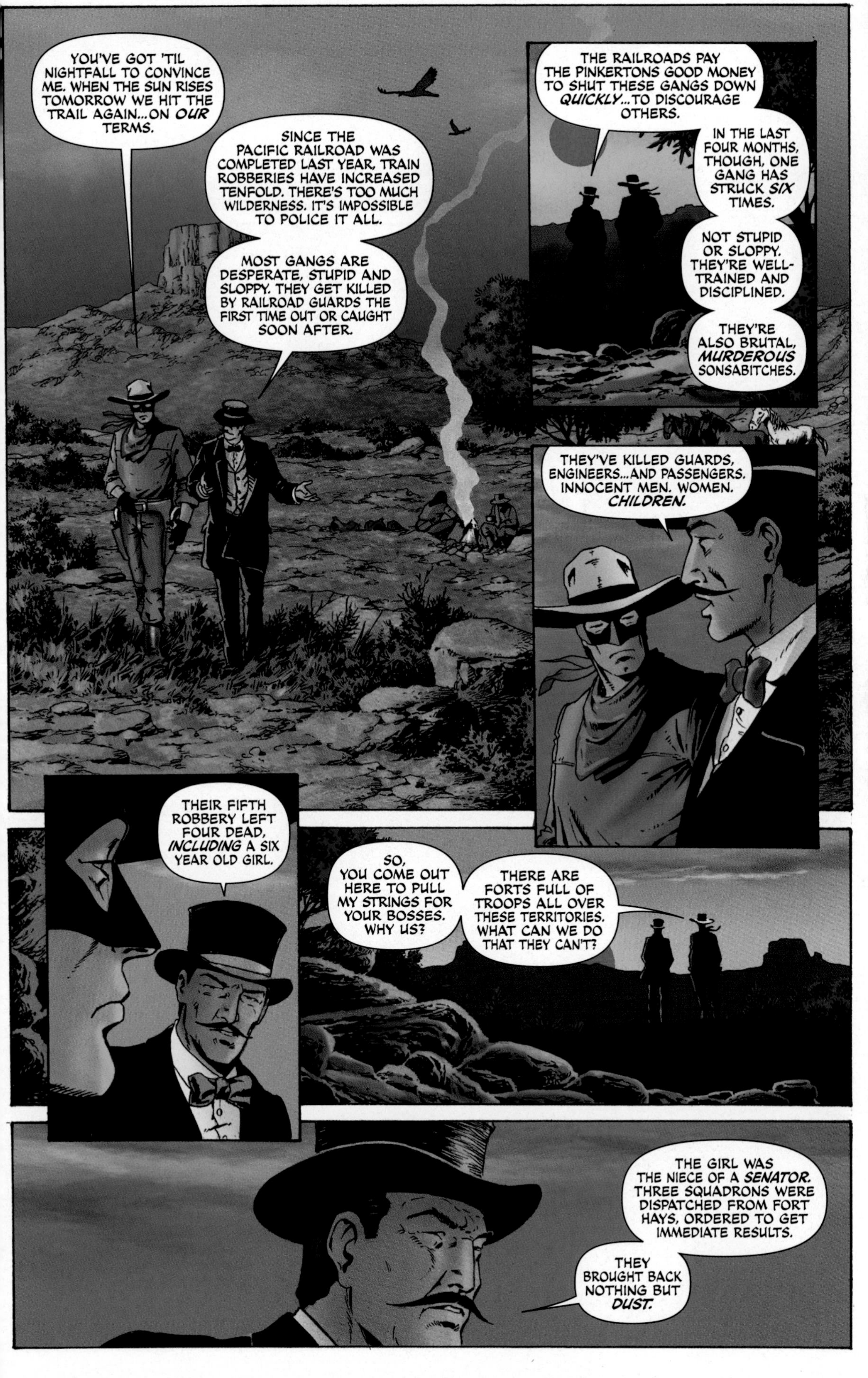
YOU'VE GOT 'TIL NIGHTFALL TO CONVINCE ME. WHEN THE SUN RISES TOMORROW WE HIT THE TRAIL AGAIN...ON *OUR* TERMS.
SINCE THE PACIFIC RAILROAD WAS COMPLETED LAST YEAR, TRAIN ROBBERIES HAVE INCREASED TENFOLD. THERE'S TOO MUCH WILDERNESS. IT'S IMPOSSIBLE TO POLICE IT ALL.
MOST GANGS ARE DESPERATE, STUPID AND SLOPPY. THEY GET KILLED BY RAILROAD GUARDS THE FIRST TIME OUT OR CAUGHT SOON AFTER.
THE RAILROADS PAY THE PINKERTONS GOOD MONEY TO SHUT THESE GANGS DOWN *QUICKLY*...TO DISCOURAGE OTHERS.
IN THE LAST FOUR MONTHS, THOUGH, ONE GANG HAS STRUCK *SIX* TIMES.
NOT STUPID OR SLOPPY. THEY'RE WELL-TRAINED AND DISCIPLINED.
THEY'RE ALSO BRUTAL, ***MURDEROUS*** SONSABITCHES.
THEY'VE KILLED GUARDS, ENGINEERS...AND PASSENGERS. INNOCENT MEN. WOMEN. ***CHILDREN.***
THEIR FIFTH ROBBERY LEFT FOUR DEAD, *INCLUDING* A SIX YEAR OLD GIRL.
SO, YOU COME OUT HERE TO PULL MY STRINGS FOR YOUR BOSSES. WHY US?
THERE ARE FORTS FULL OF TROOPS ALL OVER THESE TERRITORIES. WHAT CAN WE DO THAT THEY CAN'T?
THE GIRL WAS THE NIECE OF A *SENATOR*. THREE SQUADRONS WERE DISPATCHED FROM FORT HAYS, ORDERED TO GET IMMEDIATE RESULTS.
THEY BROUGHT BACK NOTHING BUT ***DUST.***

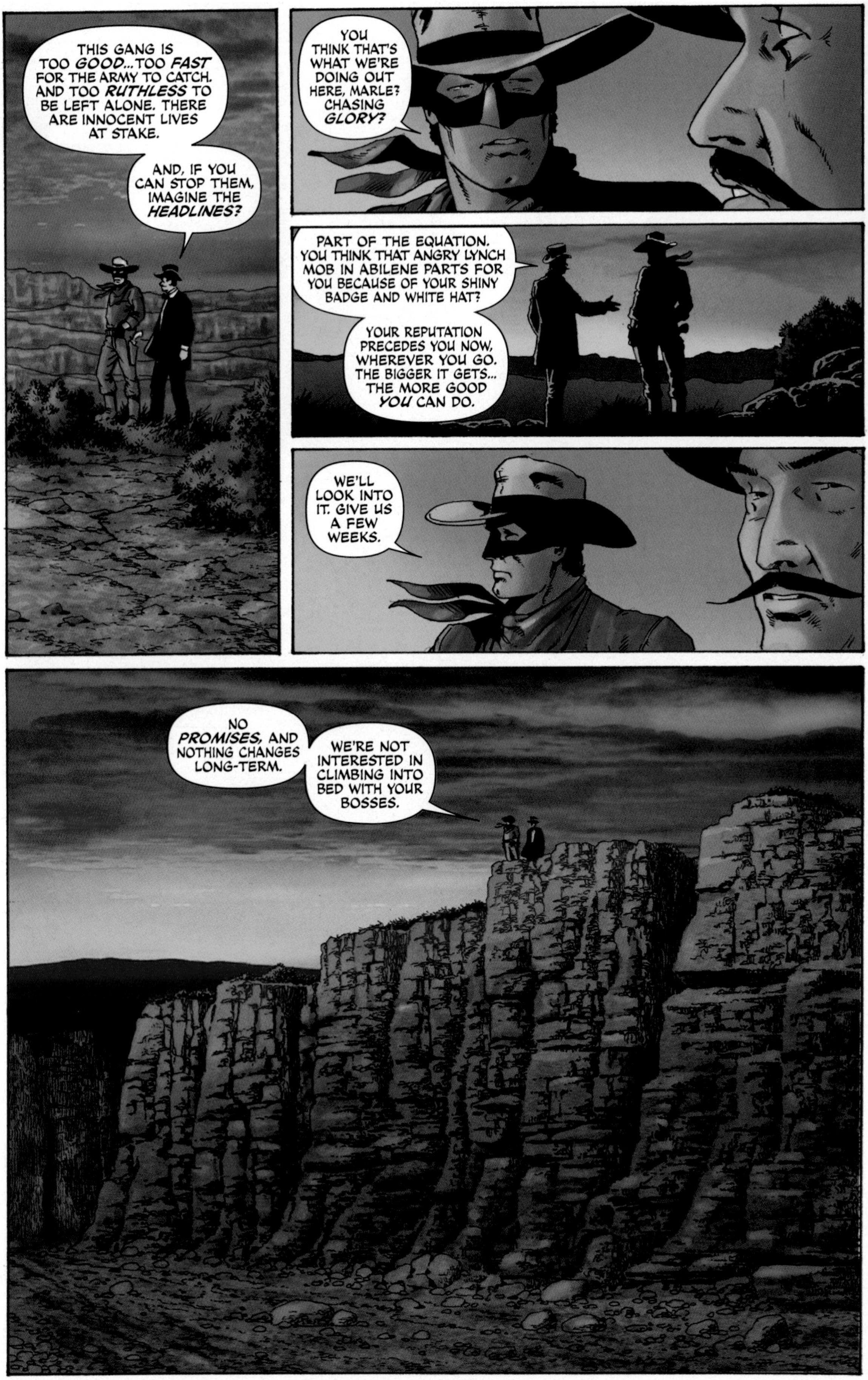
THIS GANG IS TOO GOOD...TOO FAST FOR THE ARMY TO CATCH. AND TOO RUTHLESS TO BE LEFT ALONE. THERE ARE INNOCENT LIVES AT STAKE.
AND, IF YOU CAN STOP THEM, IMAGINE THE HEADLINES?
YOU THINK THAT'S WHAT WE'RE DOING OUT HERE, MARLE? CHASING GLORY?
PART OF THE EQUATION. YOU THINK THAT ANGRY LYNCH MOB IN ABILENE PARTS FOR YOU BECAUSE OF YOUR SHINY BADGE AND WHITE HAT?
YOUR REPUTATION PRECEDES YOU NOW, WHEREVER YOU GO. THE BIGGER IT GETS... THE MORE GOOD YOU CAN DO.
WE'LL LOOK INTO IT. GIVE US A FEW WEEKS.
NO PROMISES, AND NOTHING CHANGES LONG-TERM.
WE'RE NOT INTERESTED IN CLIMBING INTO BED WITH YOUR BOSSES.

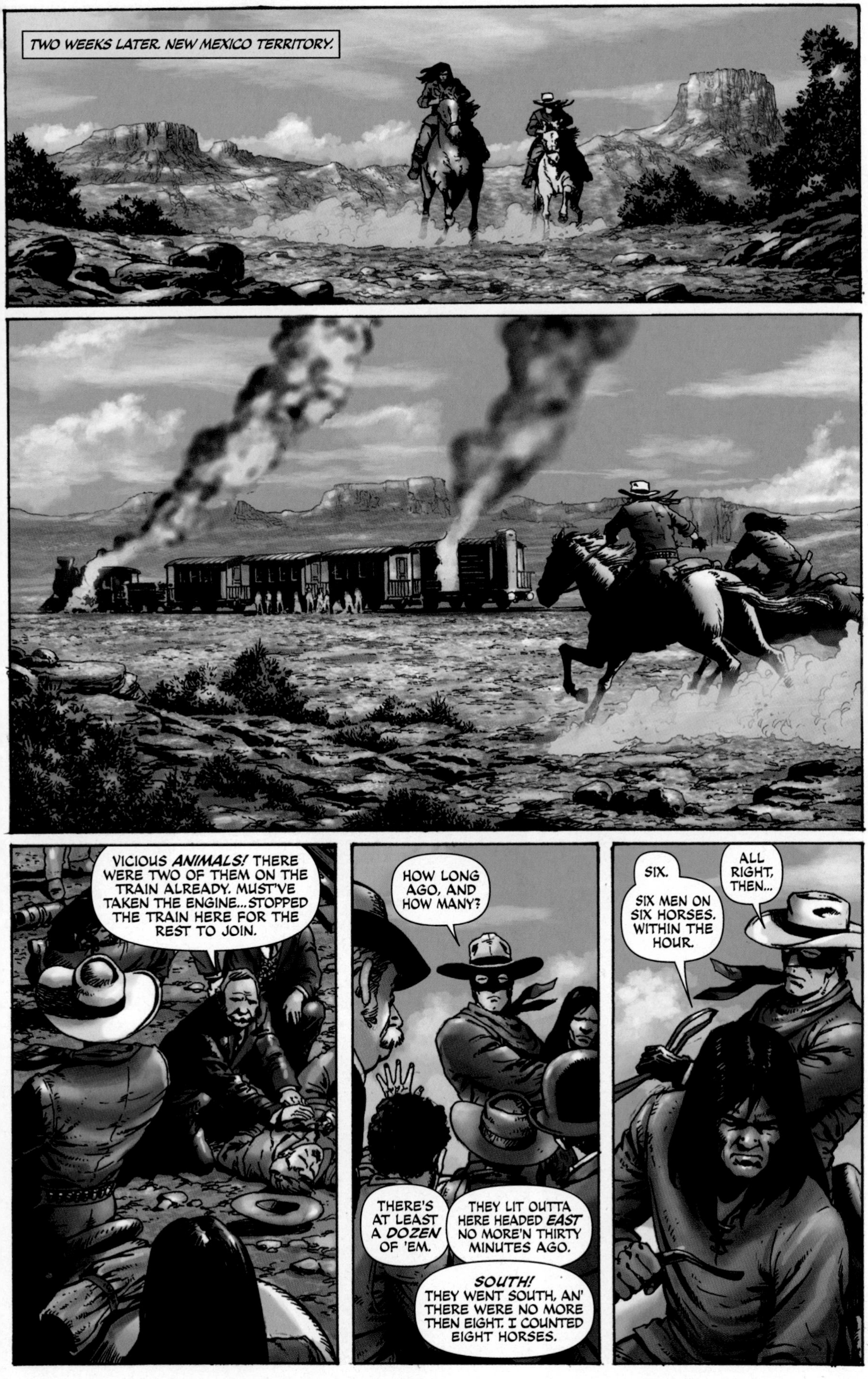
TWO WEEKS LATER. NEW MEXICO TERRITORY.
VICIOUS *ANIMALS!* THERE WERE TWO OF THEM ON THE TRAIN ALREADY. MUST'VE TAKEN THE ENGINE...STOPPED THE TRAIN HERE FOR THE REST TO JOIN.
HOW LONG AGO, AND HOW MANY?
THERE'S AT LEAST A *DOZEN* OF 'EM.
THEY LIT OUTTA HERE HEADED *EAST* NO MORE'N THIRTY MINUTES AGO.
SOUTH! THEY WENT SOUTH, AN' THERE WERE NO MORE THEN EIGHT. I COUNTED EIGHT HORSES.
SIX.
SIX MEN ON SIX HORSES. WITHIN THE HOUR.
ALL RIGHT, THEN...

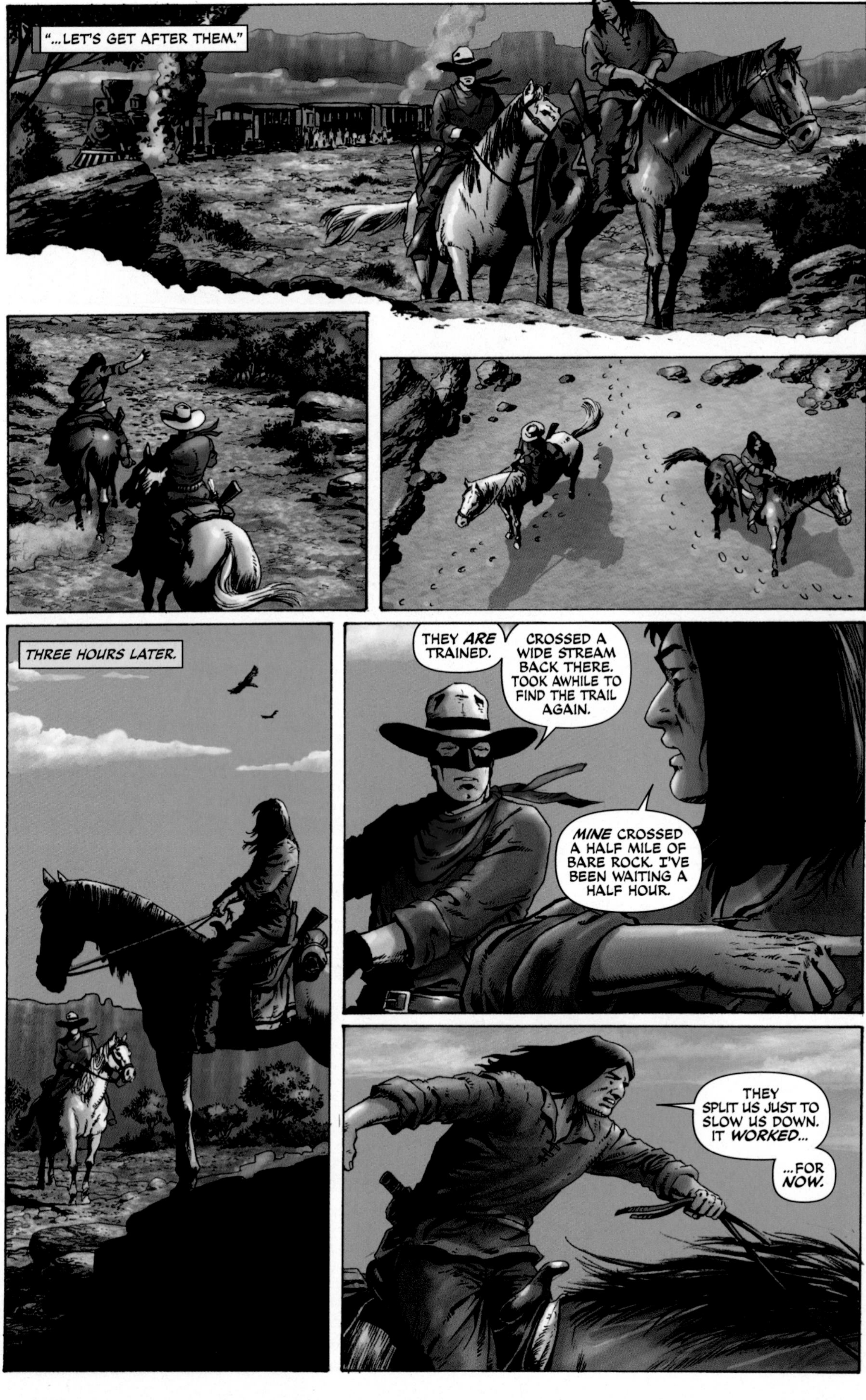
"...LET'S GET AFTER THEM."
THREE HOURS LATER.
THEY *ARE* TRAINED.
CROSSED A WIDE STREAM BACK THERE. TOOK AWHILE TO FIND THE TRAIL AGAIN.
MINE CROSSED A HALF MILE OF BARE ROCK. I'VE BEEN WAITING A HALF HOUR.
THEY SPLIT US JUST TO SLOW US DOWN. IT *WORKED*...
...FOR *NOW.*

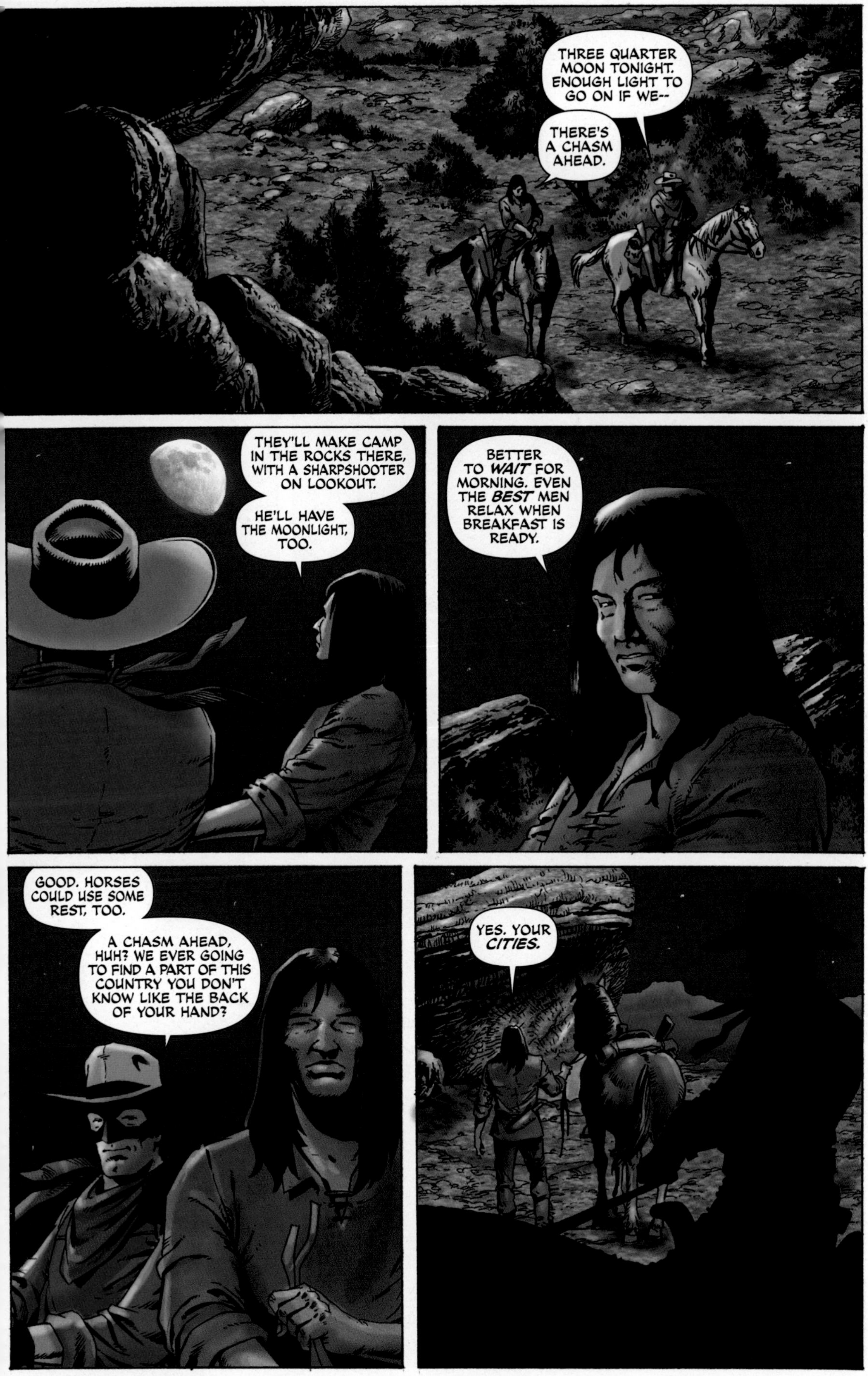
THREE QUARTER MOON TONIGHT. ENOUGH LIGHT TO GO ON IF WE--
THERE'S A CHASM AHEAD.
THEY'LL MAKE CAMP IN THE ROCKS THERE, WITH A SHARPSHOOTER ON LOOKOUT.
HE'LL HAVE THE MOONLIGHT, TOO.
BETTER TO *WAIT* FOR MORNING. EVEN THE *BEST* MEN RELAX WHEN BREAKFAST IS READY.
GOOD. HORSES COULD USE SOME REST, TOO.
A CHASM AHEAD, HUH? WE EVER GOING TO FIND A PART OF THIS COUNTRY YOU DON'T KNOW LIKE THE BACK OF YOUR HAND?
YES. YOUR *CITIES.*

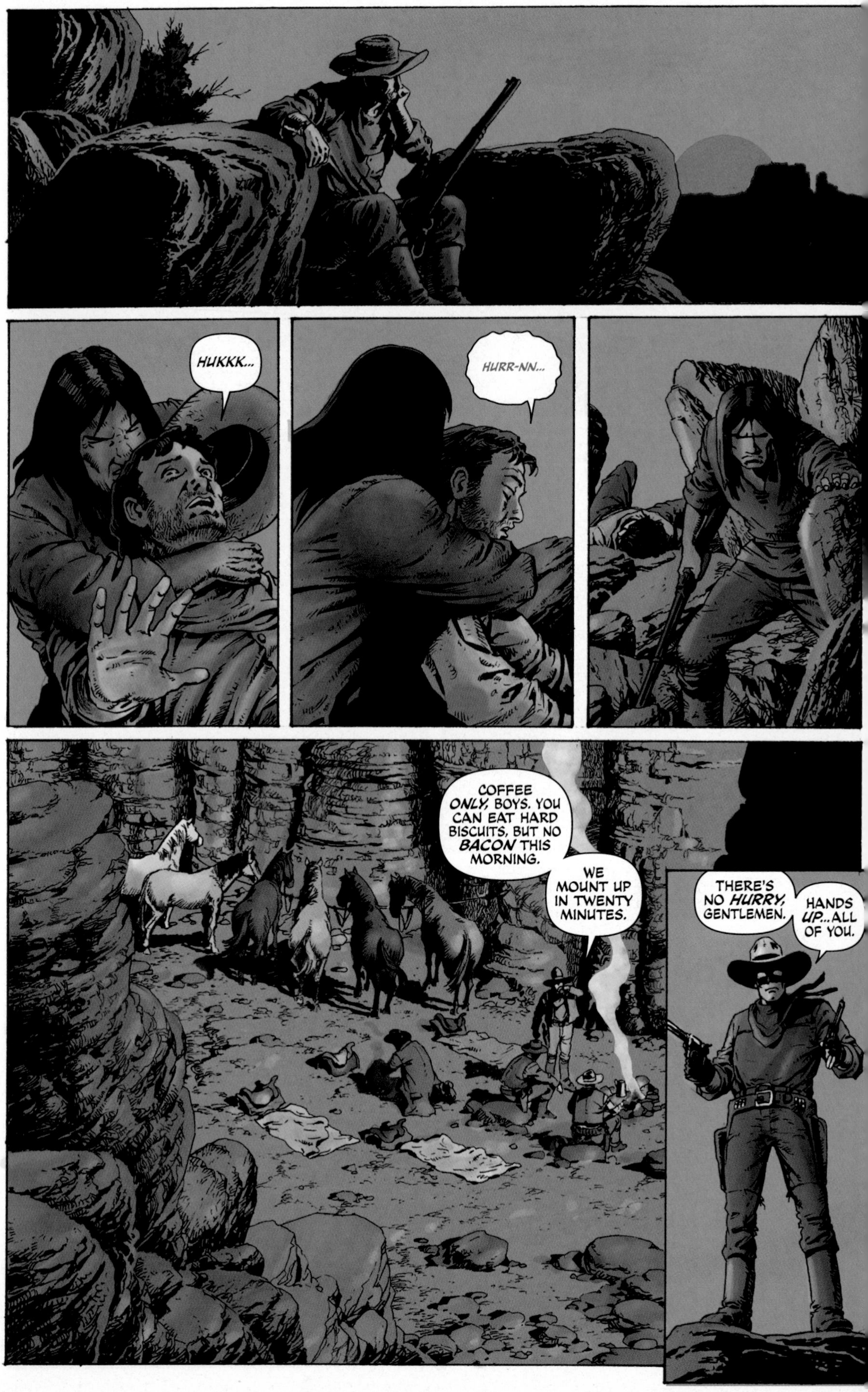
HUKKK...
HURR-NN...
COFFEE ONLY, BOYS. YOU CAN EAT HARD BISCUITS, BUT NO BACON THIS MORNING.
WE MOUNT UP IN TWENTY MINUTES.
THERE'S NO HURRY, GENTLEMEN.
HANDS UP...ALL OF YOU.

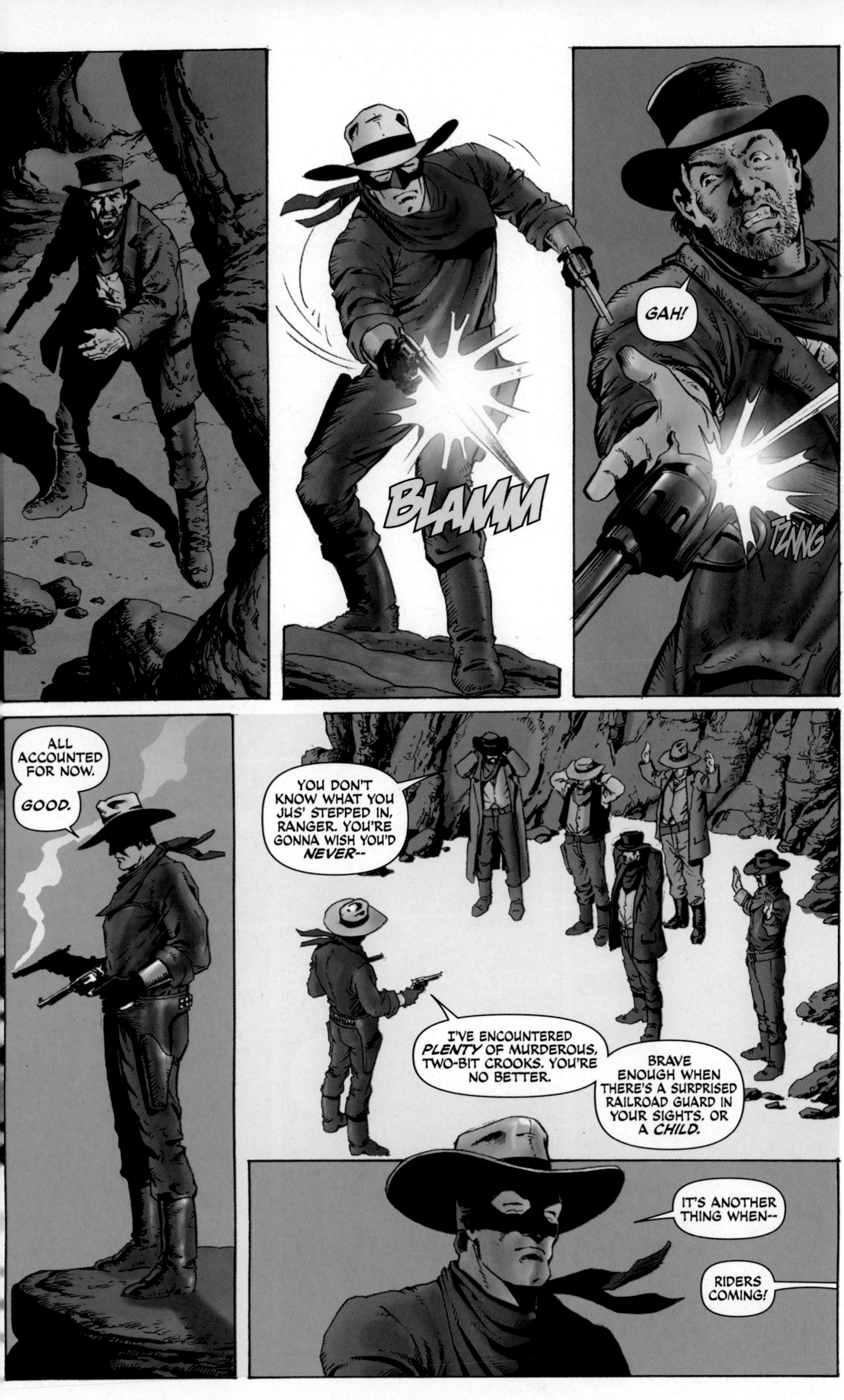
BLAMM
GAH!
TZNNG
ALL ACCOUNTED FOR NOW.
GOOD.
YOU DON'T KNOW WHAT YOU JUS' STEPPED IN, RANGER. YOU'RE GONNA WISH YOU'D *NEVER*--
I'VE ENCOUNTERED *PLENTY* OF MURDEROUS, TWO-BIT CROOKS. YOU'RE NO BETTER.
BRAVE ENOUGH WHEN THERE'S A SURPRISED RAILROAD GUARD IN YOUR SIGHTS. OR A *CHILD.*
IT'S ANOTHER THING WHEN--
RIDERS COMING!

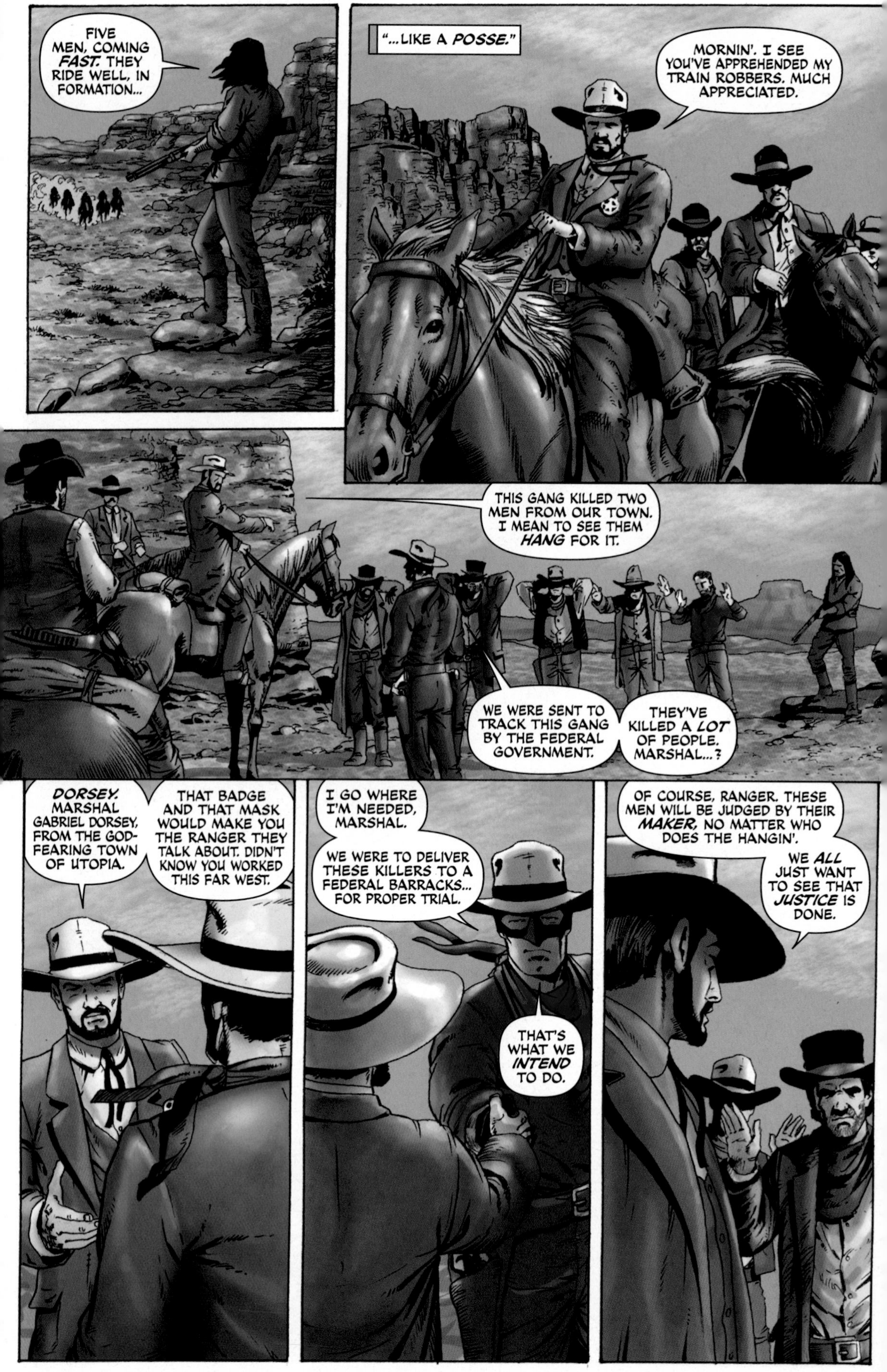
FIVE MEN, COMING *FAST.* THEY RIDE WELL, IN FORMATION...
"...LIKE A *POSSE.*"
MORNIN'. I SEE YOU'VE APPREHENDED MY TRAIN ROBBERS. MUCH APPRECIATED.
THIS GANG KILLED TWO MEN FROM OUR TOWN. I MEAN TO SEE THEM *HANG* FOR IT.
WE WERE SENT TO TRACK THIS GANG BY THE FEDERAL GOVERNMENT.
THEY'VE KILLED A *LOT* OF PEOPLE. MARSHAL...?
DORSEY. MARSHAL GABRIEL DORSEY, FROM THE GOD-FEARING TOWN OF UTOPIA.
THAT BADGE AND THAT MASK WOULD MAKE YOU THE RANGER THEY TALK ABOUT. DIDN'T KNOW YOU WORKED THIS FAR WEST.
I GO WHERE I'M NEEDED, MARSHAL.
WE WERE TO DELIVER THESE KILLERS TO A FEDERAL BARRACKS... FOR PROPER TRIAL.
THAT'S WHAT WE *INTEND* TO DO.
OF COURSE, RANGER. THESE MEN WILL BE JUDGED BY THEIR *MAKER*, NO MATTER WHO DOES THE HANGIN'.
WE *ALL* JUST WANT TO SEE THAT *JUSTICE* IS DONE.

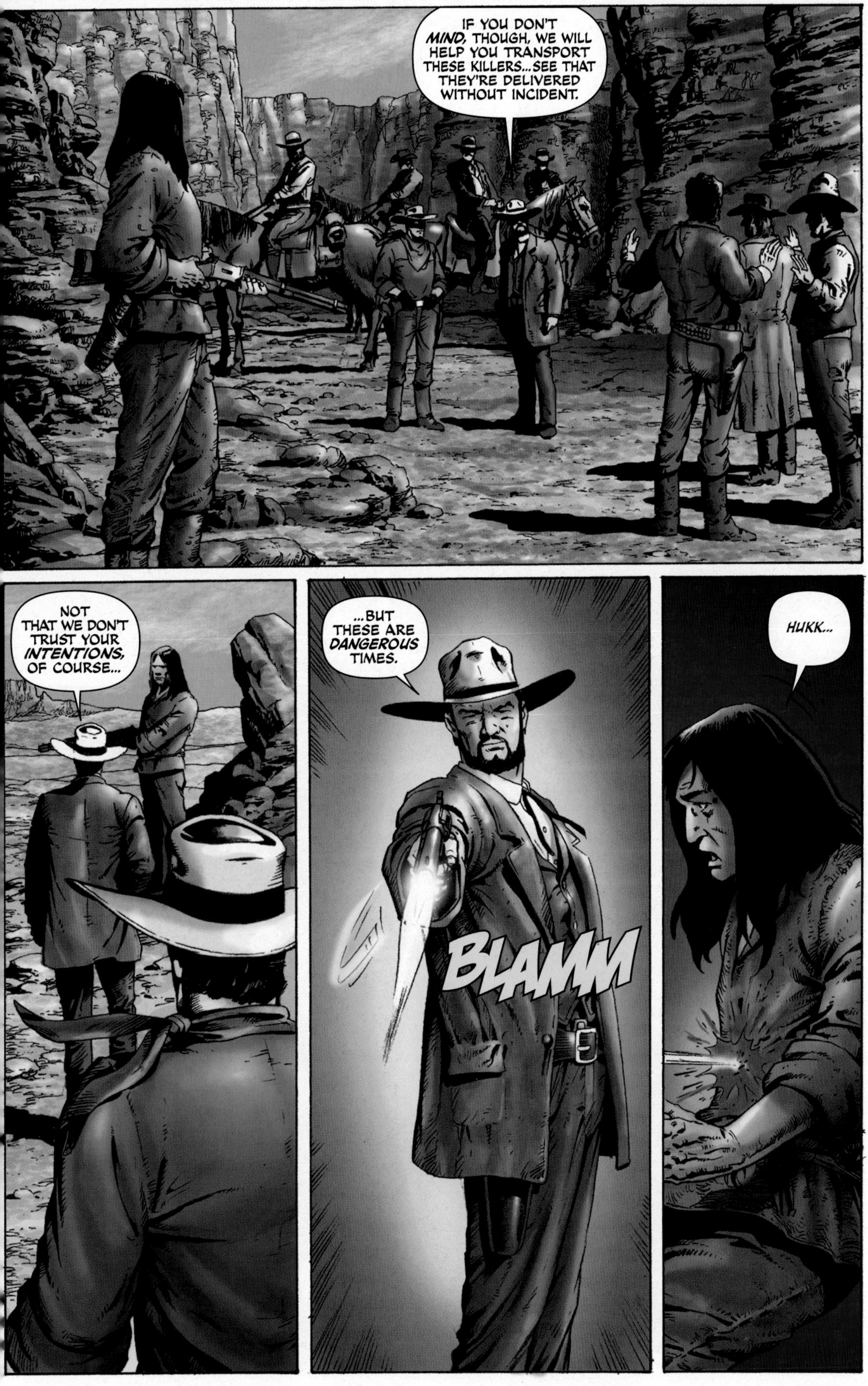
IF YOU DON'T *MIND*, THOUGH, WE WILL HELP YOU TRANSPORT THESE KILLERS...SEE THAT THEY'RE DELIVERED WITHOUT INCIDENT.
NOT THAT WE DON'T TRUST YOUR *INTENTIONS*, OF COURSE...
...BUT THESE ARE *DANGEROUS* TIMES.
BLAMM
HUKK...

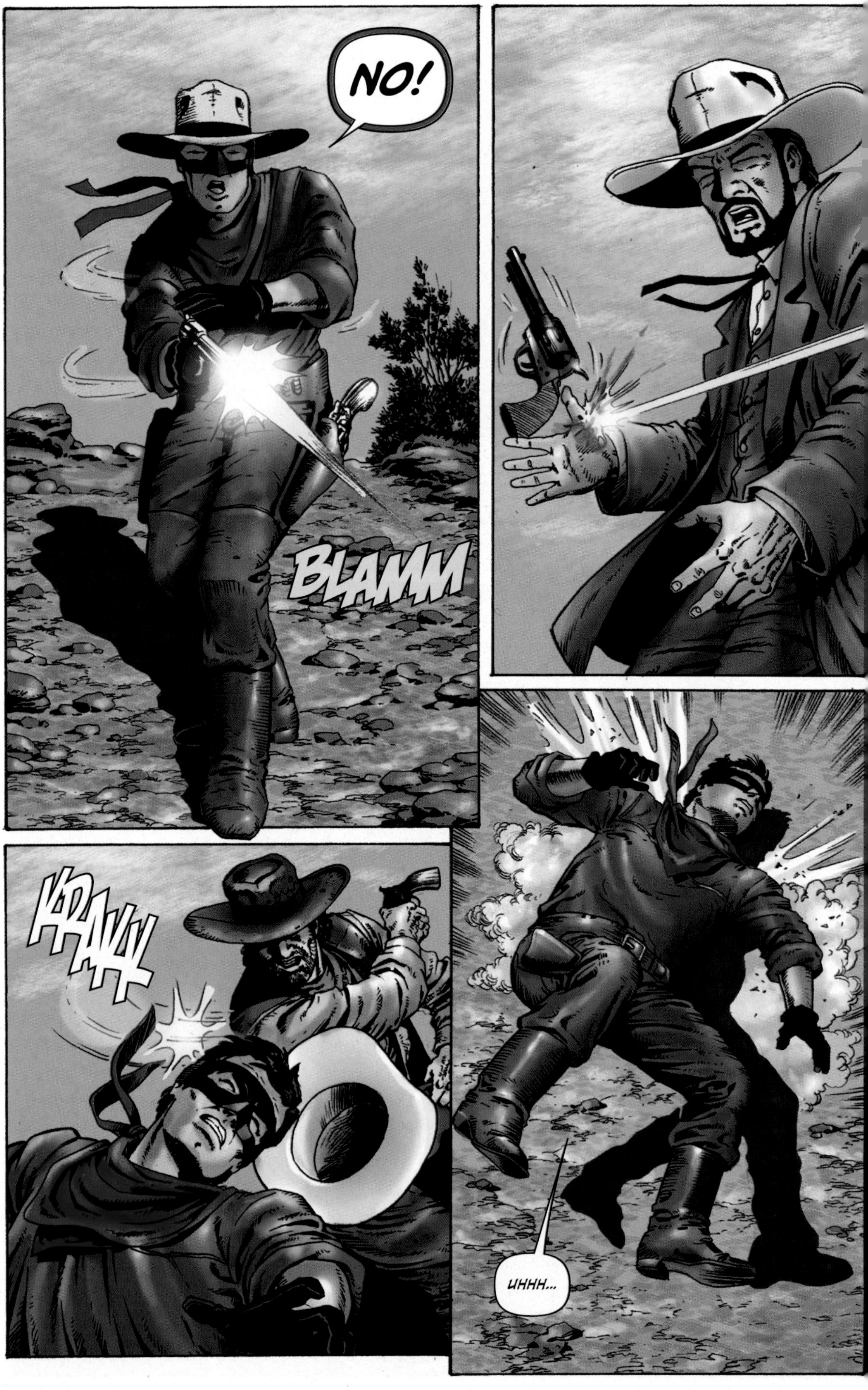
NO!
BLAMM
KRAKK
UHHH...

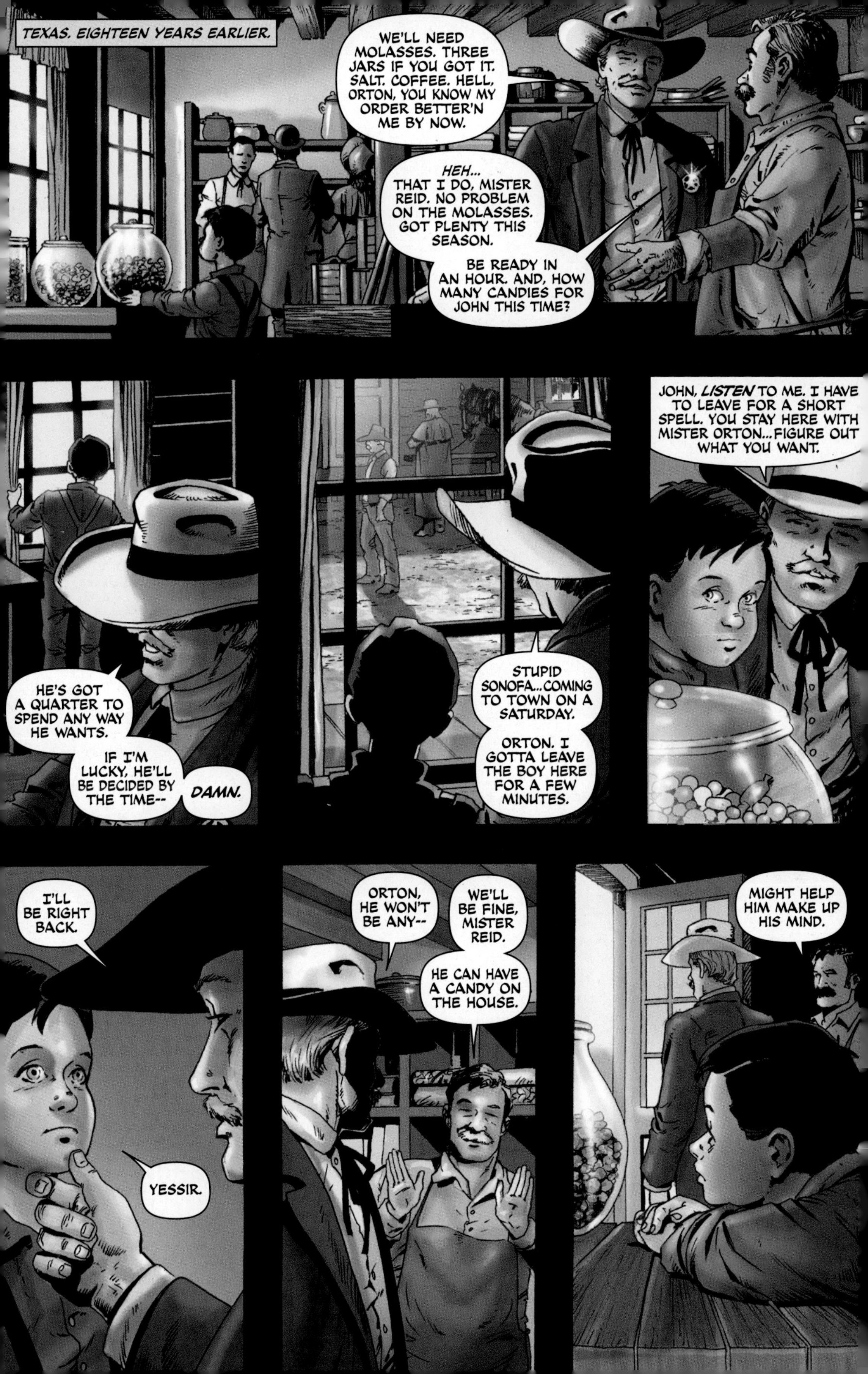
TEXAS. EIGHTEEN YEARS EARLIER.
WE'LL NEED MOLASSES. THREE JARS IF YOU GOT IT. SALT. COFFEE. HELL, ORTON, YOU KNOW MY ORDER BETTER'N ME BY NOW.
HEH... THAT I DO, MISTER REID. NO PROBLEM ON THE MOLASSES. GOT PLENTY THIS SEASON.
BE READY IN AN HOUR. AND, HOW MANY CANDIES FOR JOHN THIS TIME?
HE'S GOT A QUARTER TO SPEND ANY WAY HE WANTS.
IF I'M LUCKY, HE'LL BE DECIDED BY THE TIME--
DAMN.
STUPID SONOFA...COMING TO TOWN ON A SATURDAY.
ORTON. I GOTTA LEAVE THE BOY HERE FOR A FEW MINUTES.
JOHN, LISTEN TO ME. I HAVE TO LEAVE FOR A SHORT SPELL. YOU STAY HERE WITH MISTER ORTON...FIGURE OUT WHAT YOU WANT.
I'LL BE RIGHT BACK.
YESSIR.
ORTON, HE WON'T BE ANY--
WE'LL BE FINE, MISTER REID.
HE CAN HAVE A CANDY ON THE HOUSE.
MIGHT HELP HIM MAKE UP HIS MIND.

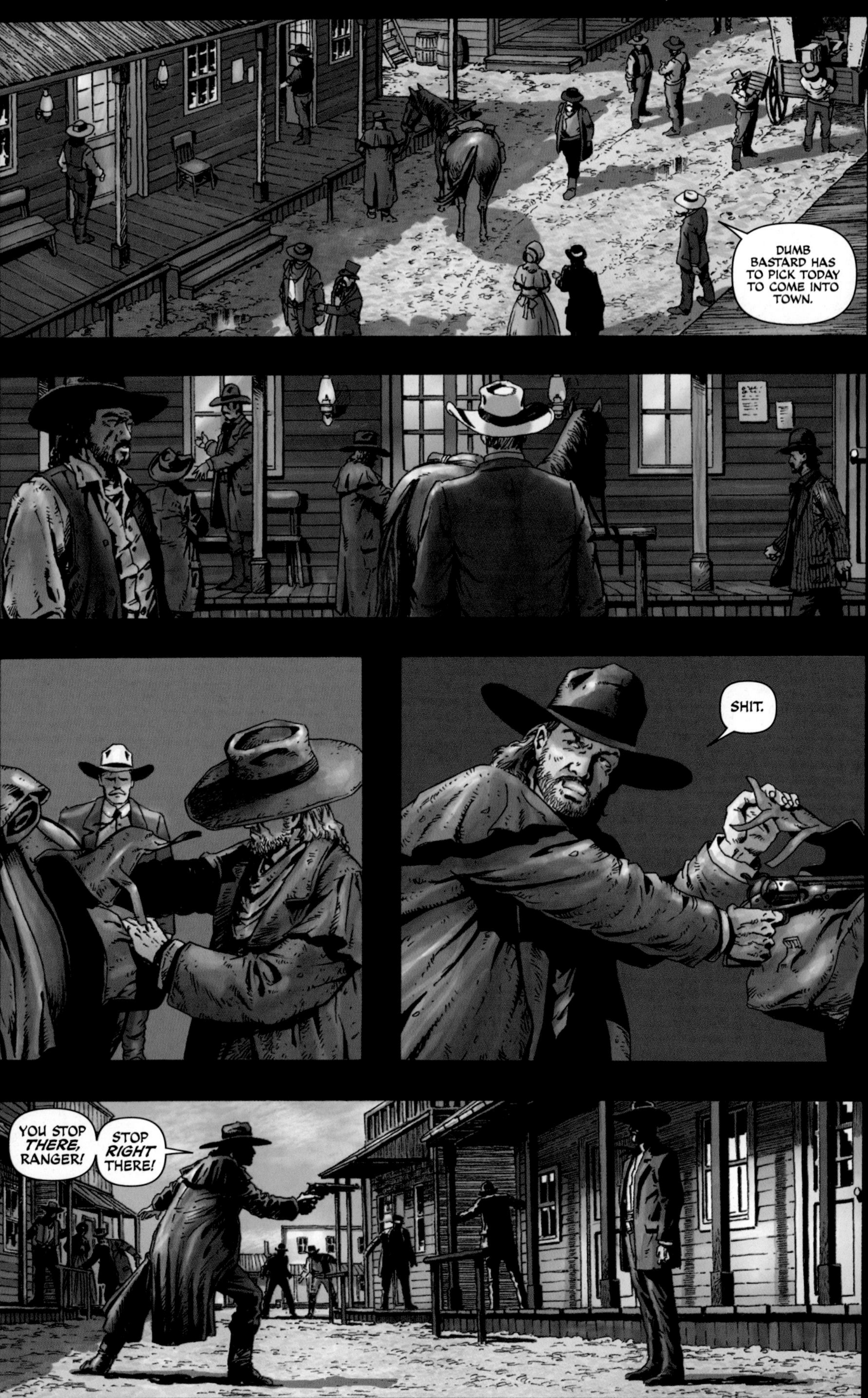
DUMB BASTARD HAS TO PICK TODAY TO COME INTO TOWN.
SHIT.
YOU STOP *THERE,* RANGER!
STOP *RIGHT* THERE!

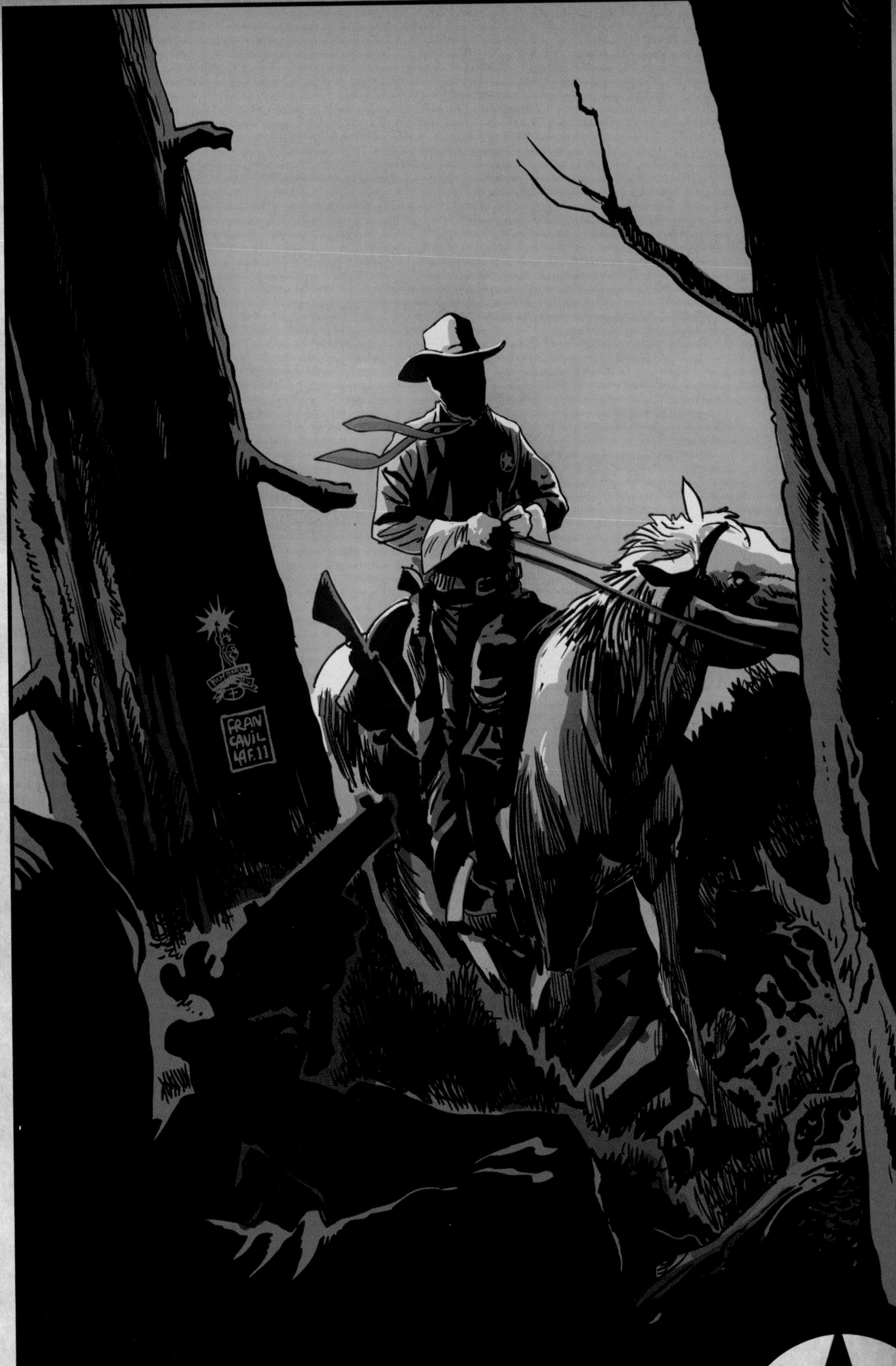

COVER BY **FRANCESCO FRANCAVILLA**

4

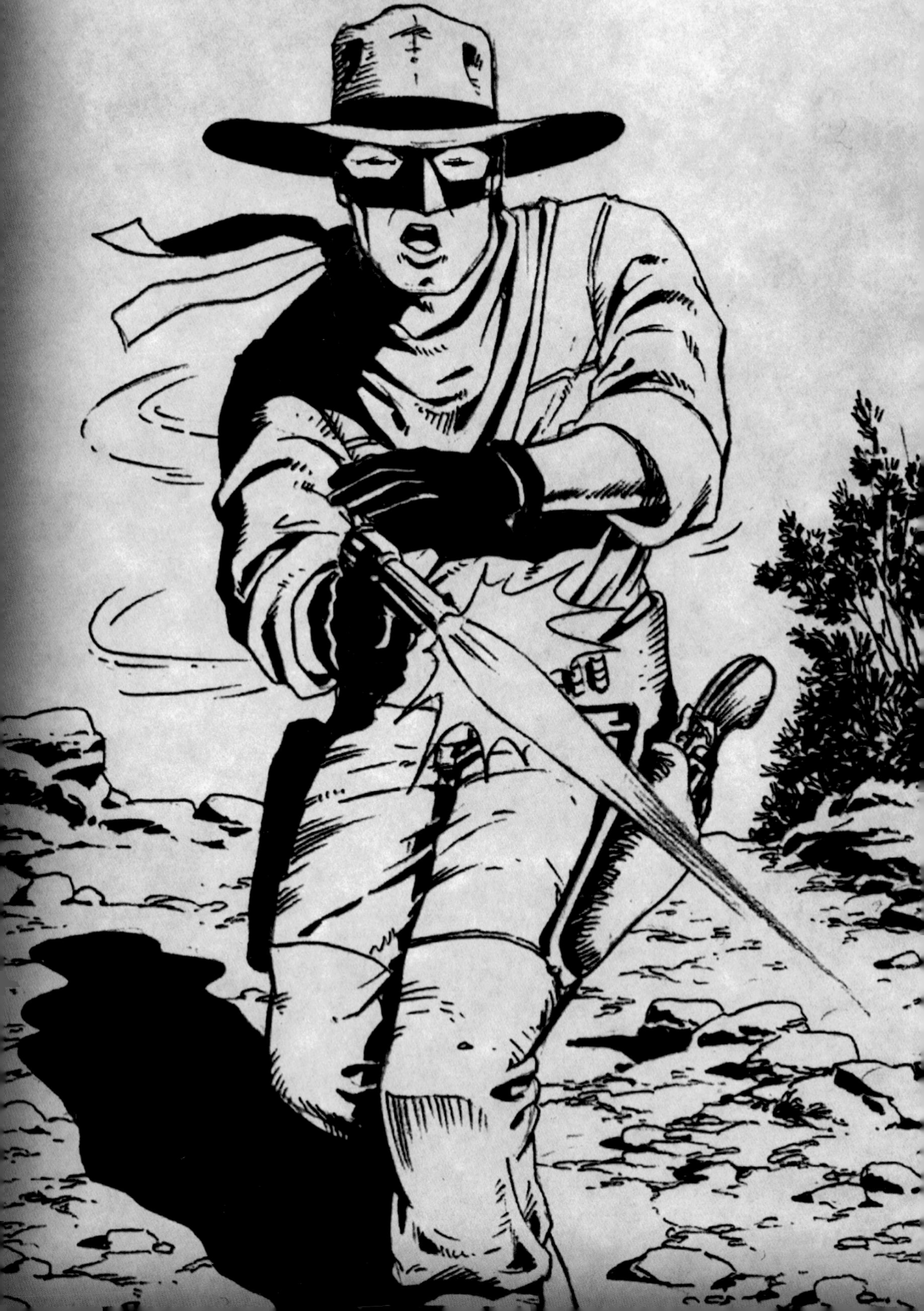

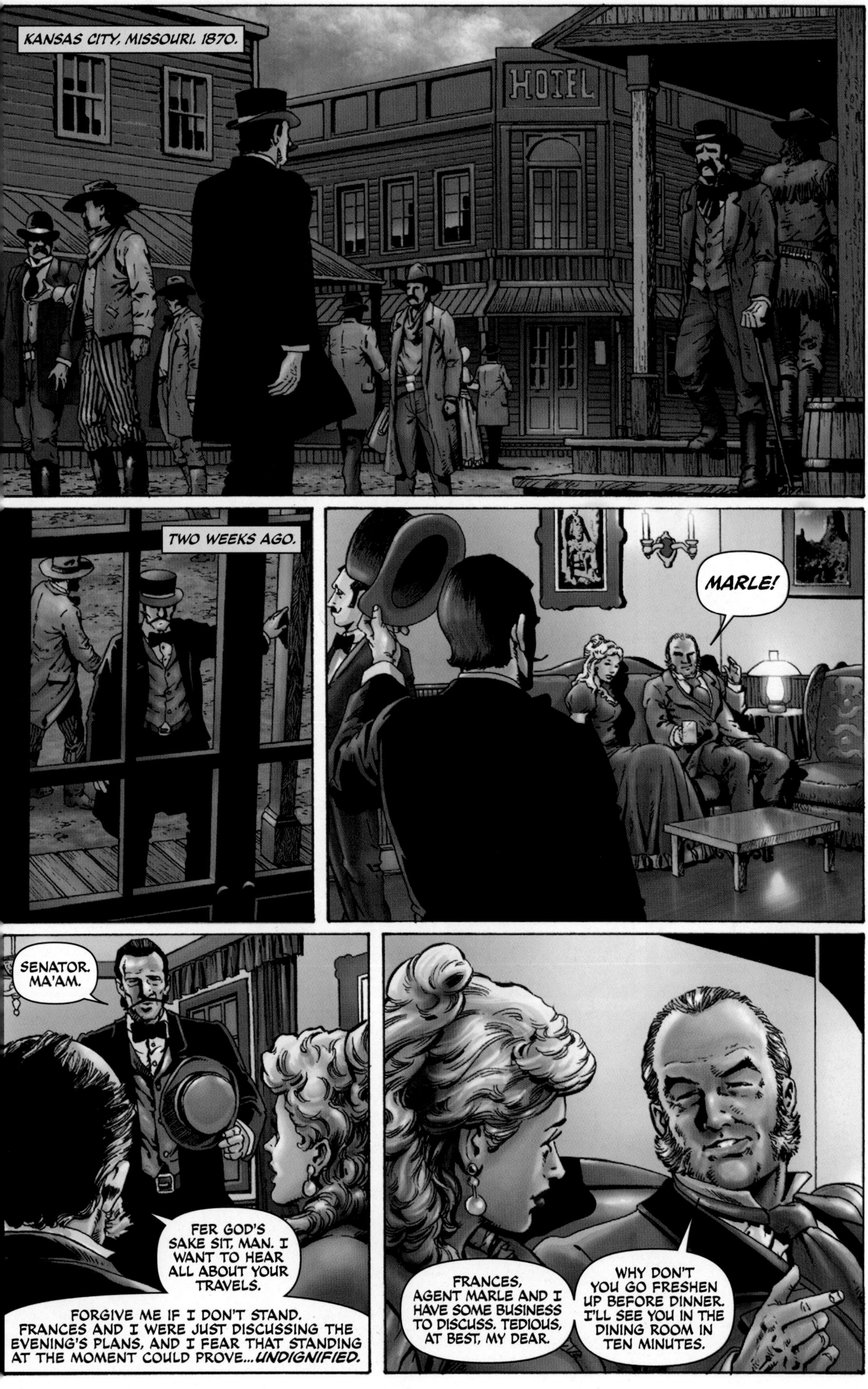
KANSAS CITY, MISSOURI. 1870.
HOTEL
TWO WEEKS AGO.
MARLE!
SENATOR. MA'AM.
FER GOD'S SAKE SIT, MAN. I WANT TO HEAR ALL ABOUT YOUR TRAVELS.
FORGIVE ME IF I DON'T STAND. FRANCES AND I WERE JUST DISCUSSING THE EVENING'S PLANS, AND I FEAR THAT STANDING AT THE MOMENT COULD PROVE...UNDIGNIFIED.
FRANCES, AGENT MARLE AND I HAVE SOME BUSINESS TO DISCUSS. TEDIOUS, AT BEST, MY DEAR.
WHY DON'T YOU GO FRESHEN UP BEFORE DINNER. I'LL SEE YOU IN THE DINING ROOM IN TEN MINUTES.

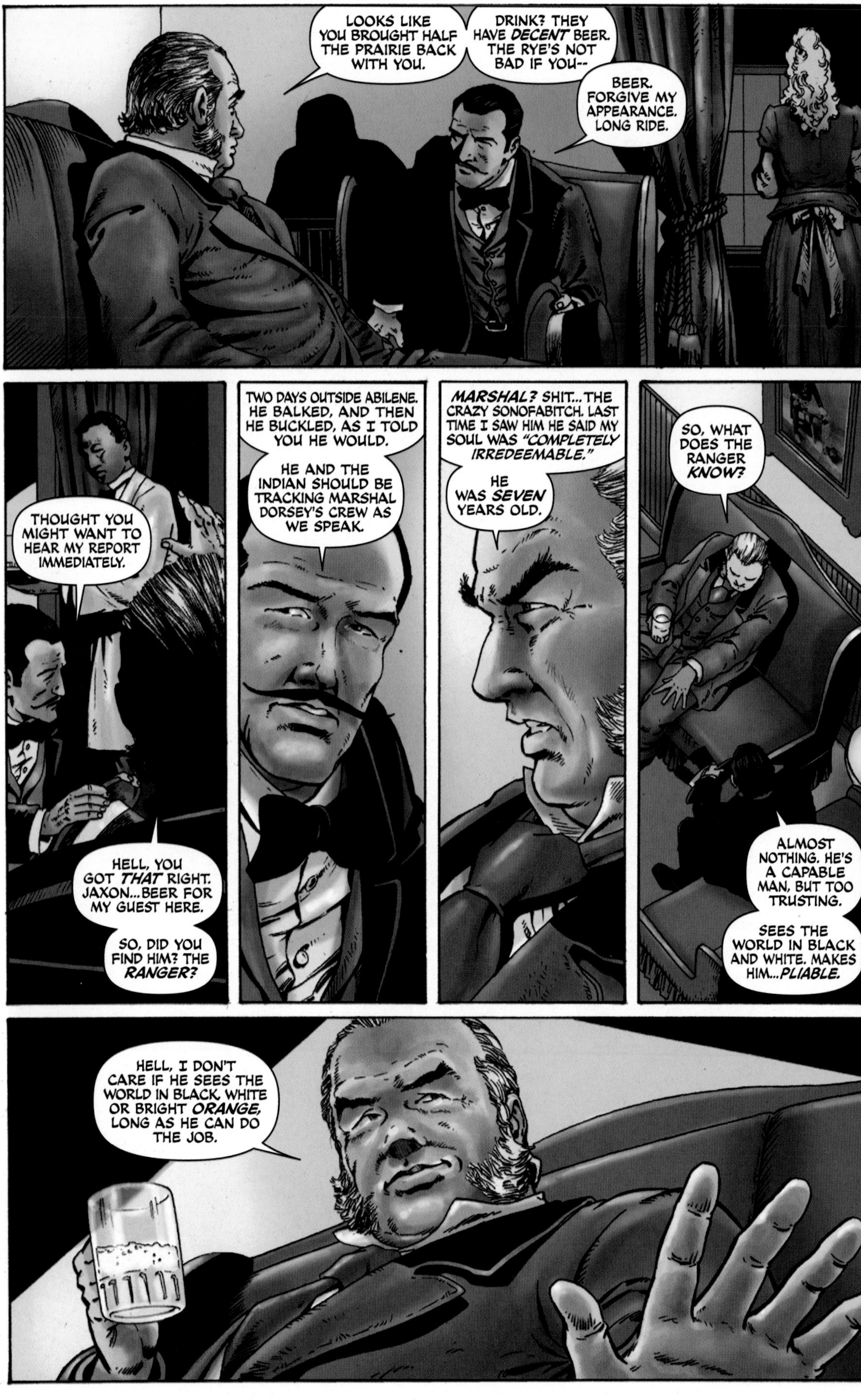
LOOKS LIKE YOU BROUGHT HALF THE PRAIRIE BACK WITH YOU.
DRINK? THEY HAVE *DECENT* BEER. THE RYE'S NOT BAD IF YOU--
BEER. FORGIVE MY APPEARANCE. LONG RIDE.
THOUGHT YOU MIGHT WANT TO HEAR MY REPORT IMMEDIATELY.
HELL, YOU GOT *THAT* RIGHT. JAXON...BEER FOR MY GUEST HERE.
SO, DID YOU FIND HIM? THE *RANGER?*
TWO DAYS OUTSIDE ABILENE. HE BALKED, AND THEN HE BUCKLED, AS I TOLD YOU HE WOULD.
HE AND THE INDIAN SHOULD BE TRACKING MARSHAL DORSEY'S CREW AS WE SPEAK.
MARSHAL? SHIT... THE CRAZY SONOFABITCH. LAST TIME I SAW HIM HE SAID MY SOUL WAS *"COMPLETELY IRREDEEMABLE."*
HE WAS *SEVEN* YEARS OLD.
SO, WHAT DOES THE RANGER *KNOW?*
ALMOST NOTHING. HE'S A CAPABLE MAN, BUT TOO TRUSTING.
SEES THE WORLD IN BLACK AND WHITE. MAKES HIM...*PLIABLE.*
HELL, I DON'T CARE IF HE SEES THE WORLD IN BLACK, WHITE OR BRIGHT *ORANGE,* LONG AS HE CAN DO THE JOB.

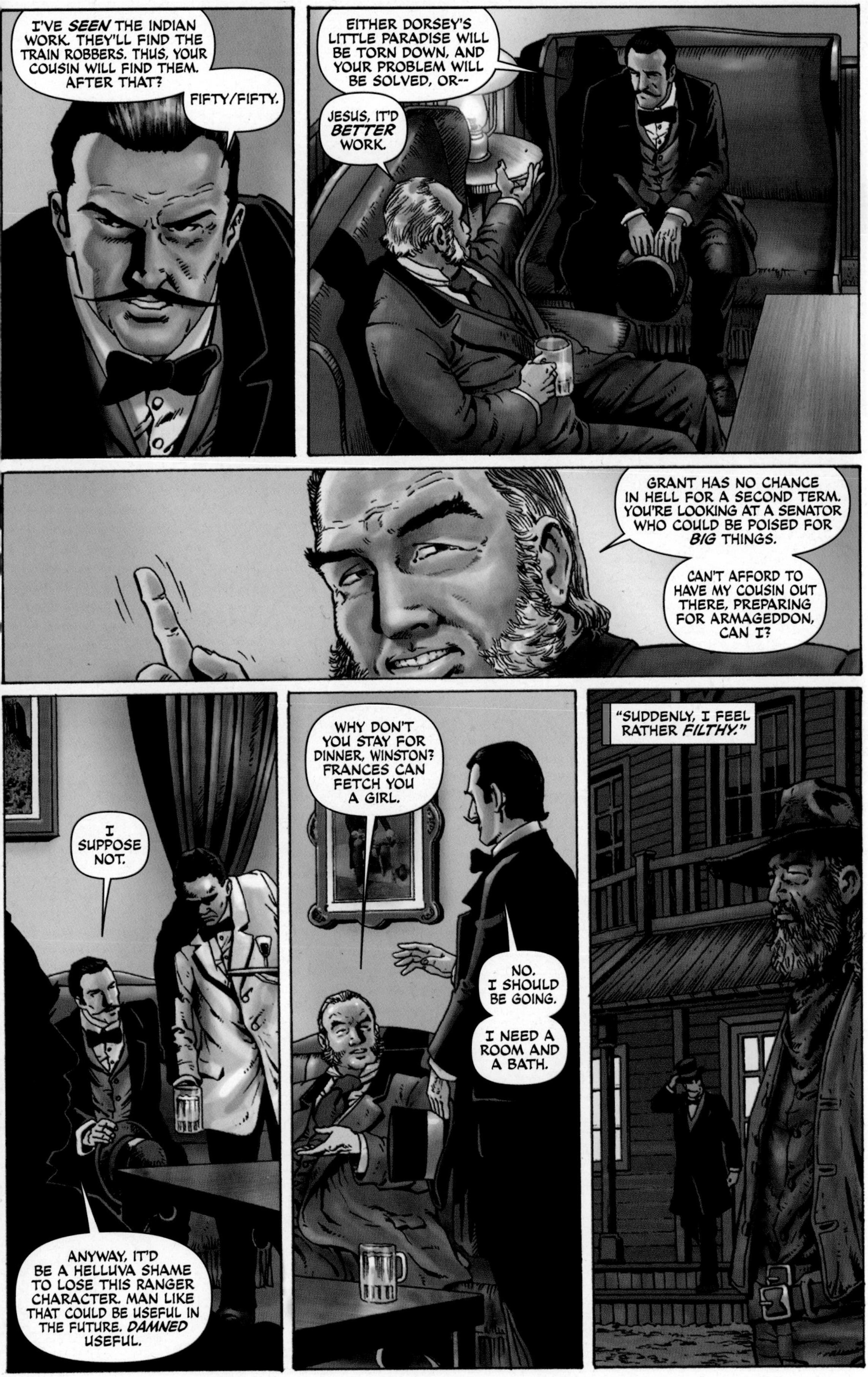
I'VE *SEEN* THE INDIAN WORK. THEY'LL FIND THE TRAIN ROBBERS. THUS, YOUR COUSIN WILL FIND THEM. AFTER THAT?
FIFTY/FIFTY.
EITHER DORSEY'S LITTLE PARADISE WILL BE TORN DOWN, AND YOUR PROBLEM WILL BE SOLVED, OR--
JESUS, IT'D *BETTER* WORK.
GRANT HAS NO CHANCE IN HELL FOR A SECOND TERM. YOU'RE LOOKING AT A SENATOR WHO COULD BE POISED FOR *BIG* THINGS.
CAN'T AFFORD TO HAVE MY COUSIN OUT THERE, PREPARING FOR ARMAGEDDON, CAN I?
I SUPPOSE NOT.
ANYWAY, IT'D BE A HELLUVA SHAME TO LOSE THIS RANGER CHARACTER. MAN LIKE THAT COULD BE USEFUL IN THE FUTURE. *DAMNED* USEFUL.
WHY DON'T YOU STAY FOR DINNER, WINSTON? FRANCES CAN FETCH YOU A GIRL.
NO. I SHOULD BE GOING.
I NEED A ROOM AND A BATH.
"SUDDENLY, I FEEL RATHER *FILTHY.*"

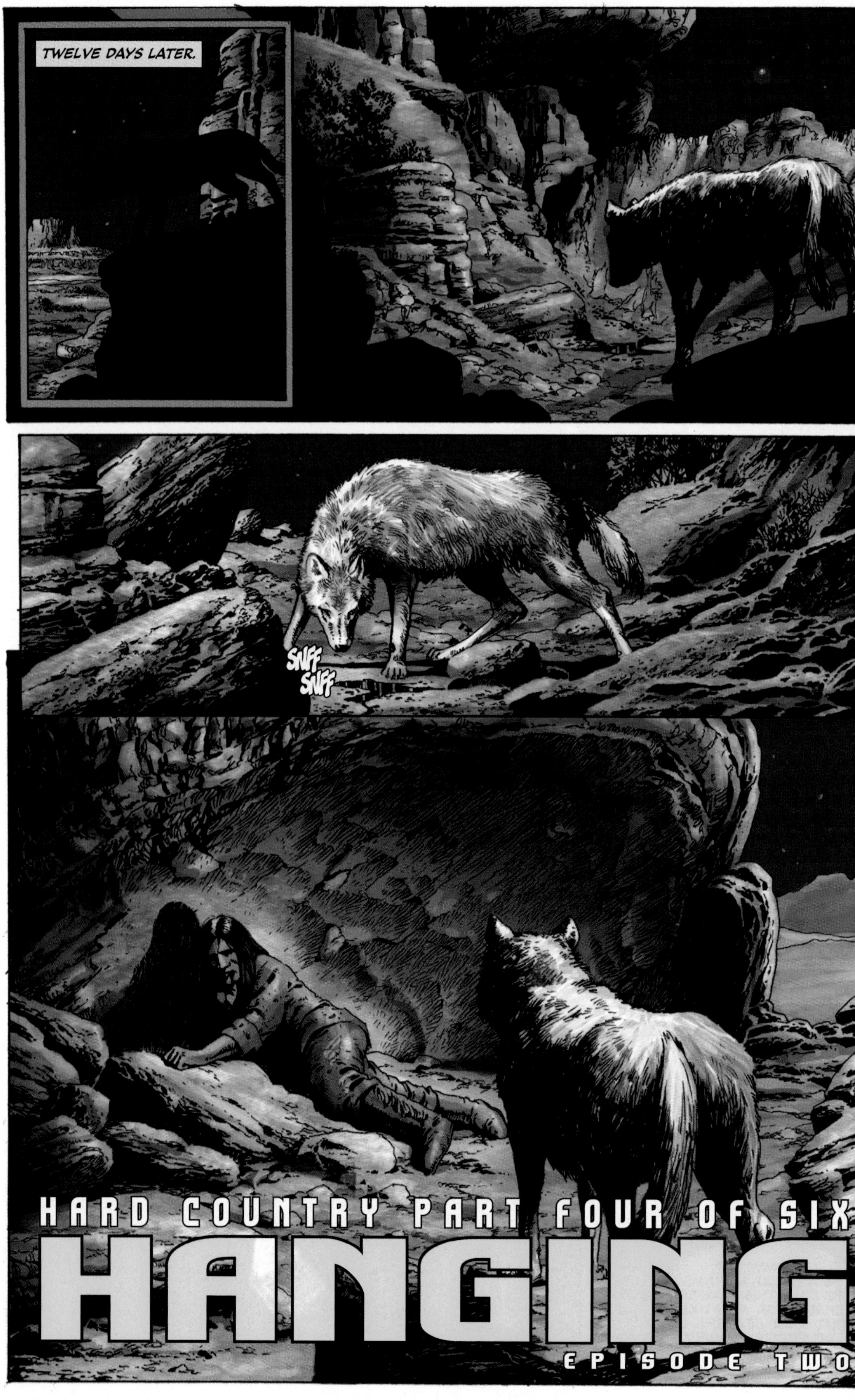
TWELVE DAYS LATER.
SNIFF
SNIFF
HARD COUNTRY PART FOUR OF SIX
HANGING
EPISODE TWO

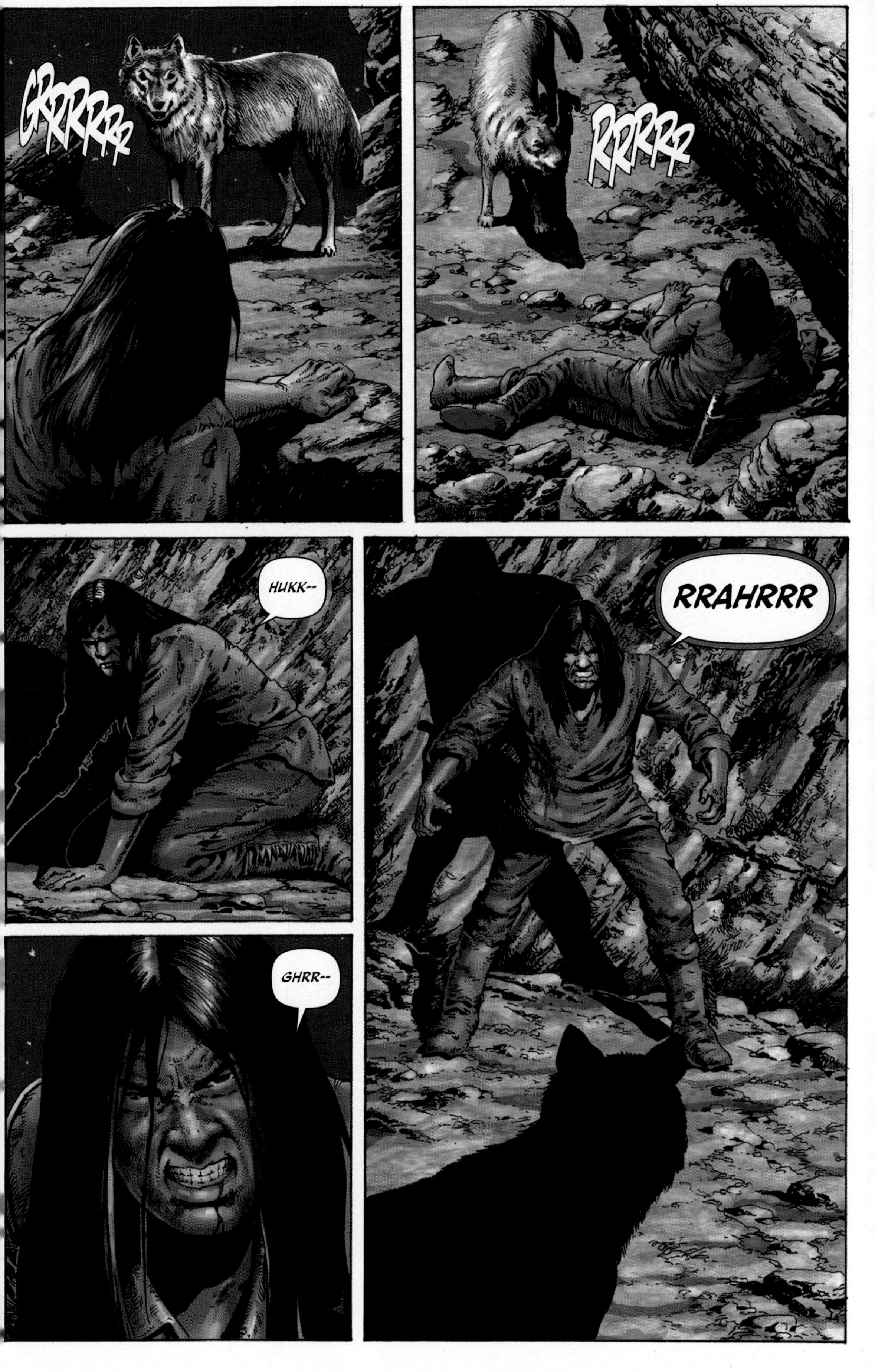
GRRRRRR
RRRRR
HUKK--
GHRR--
RRAHRRR

RRRRR

≥HUHHG≥
≥HUHNN≥

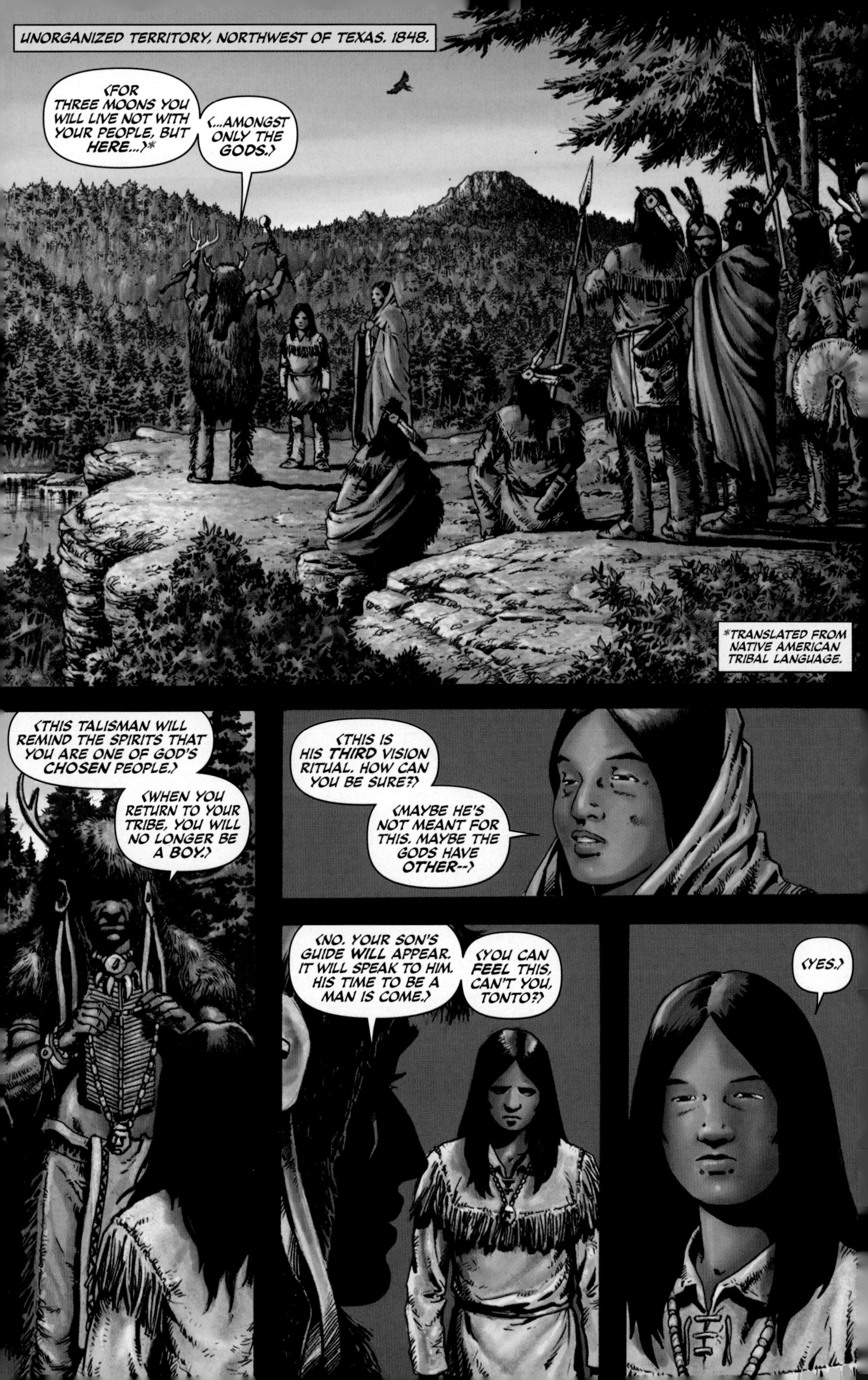
UNORGANIZED TERRITORY, NORTHWEST OF TEXAS. 1848.
‹FOR THREE MOONS YOU WILL LIVE NOT WITH YOUR PEOPLE, BUT HERE...›*
‹...AMONGST ONLY THE GODS.›
*TRANSLATED FROM NATIVE AMERICAN TRIBAL LANGUAGE.
‹THIS TALISMAN WILL REMIND THE SPIRITS THAT YOU ARE ONE OF GOD'S CHOSEN PEOPLE.›
‹WHEN YOU RETURN TO YOUR TRIBE, YOU WILL NO LONGER BE A BOY.›
‹THIS IS HIS THIRD VISION RITUAL. HOW CAN YOU BE SURE?›
‹MAYBE HE'S NOT MEANT FOR THIS. MAYBE THE GODS HAVE OTHER--›
‹NO. YOUR SON'S GUIDE WILL APPEAR. IT WILL SPEAK TO HIM. HIS TIME TO BE A MAN IS COME.›
‹YOU CAN FEEL THIS, CAN'T YOU, TONTO?›
‹YES.›

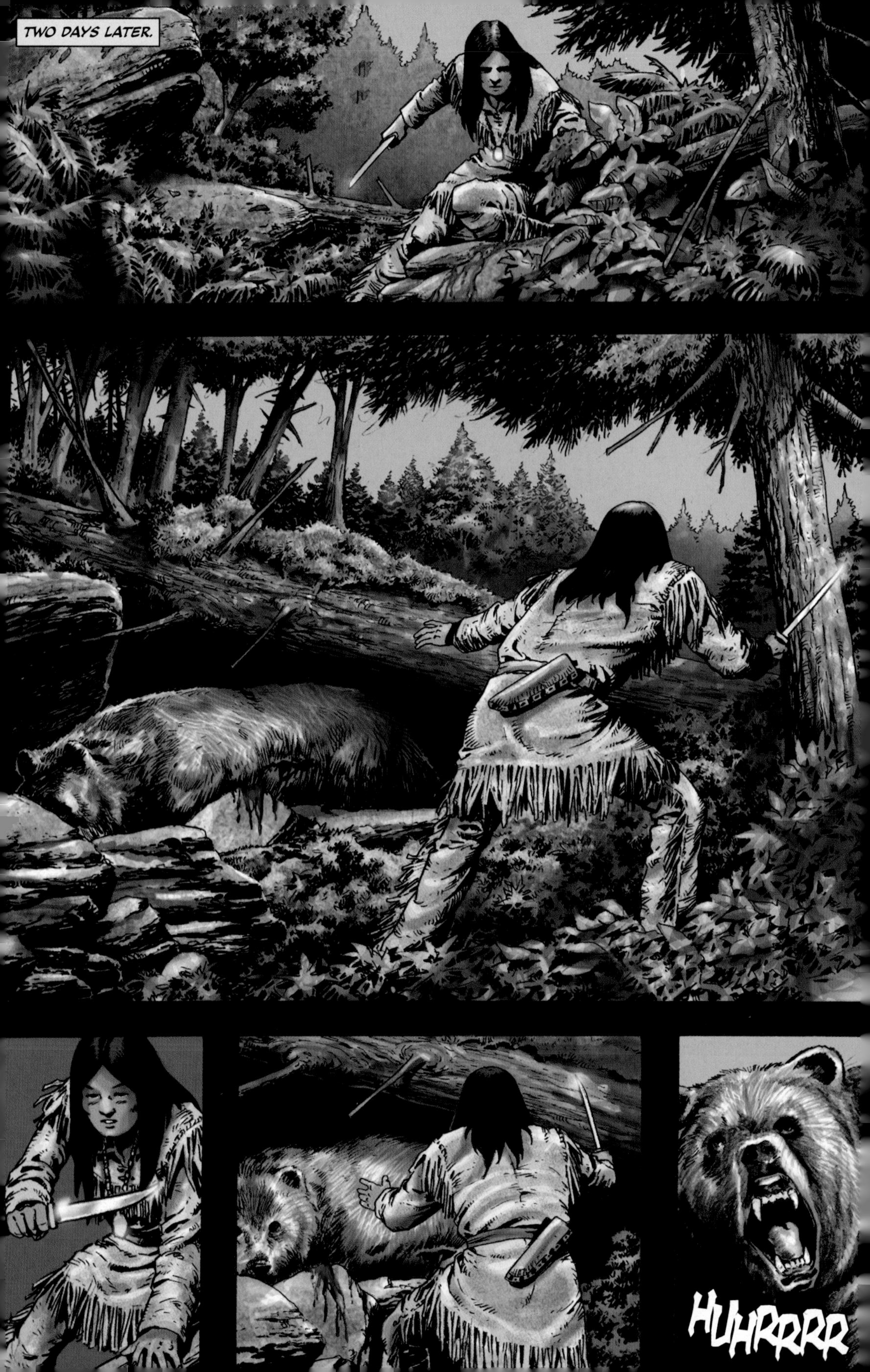
TWO DAYS LATER.
HUHRRRR

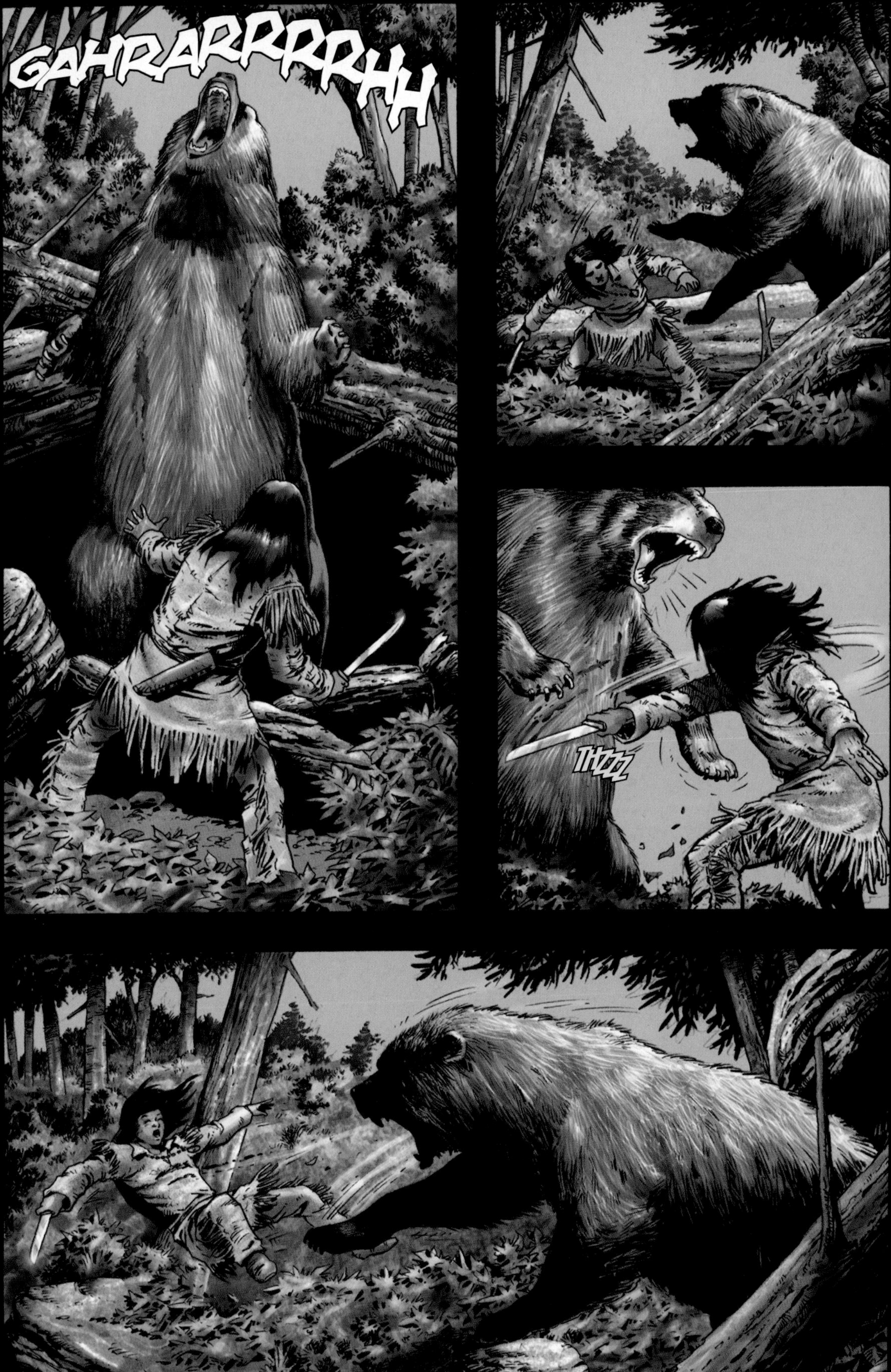
GAHRARRRRRHH
THZZZ

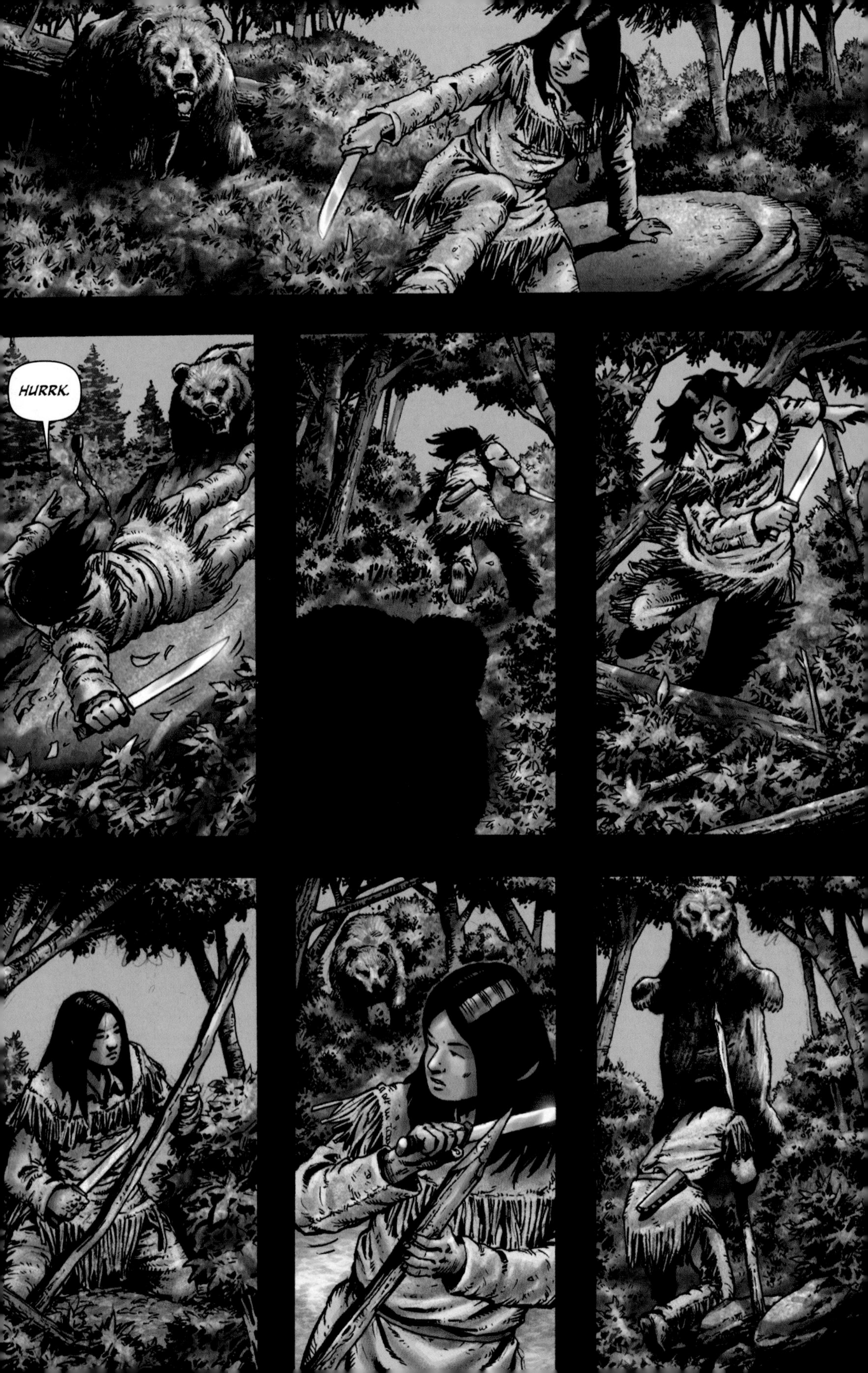
HURRK.

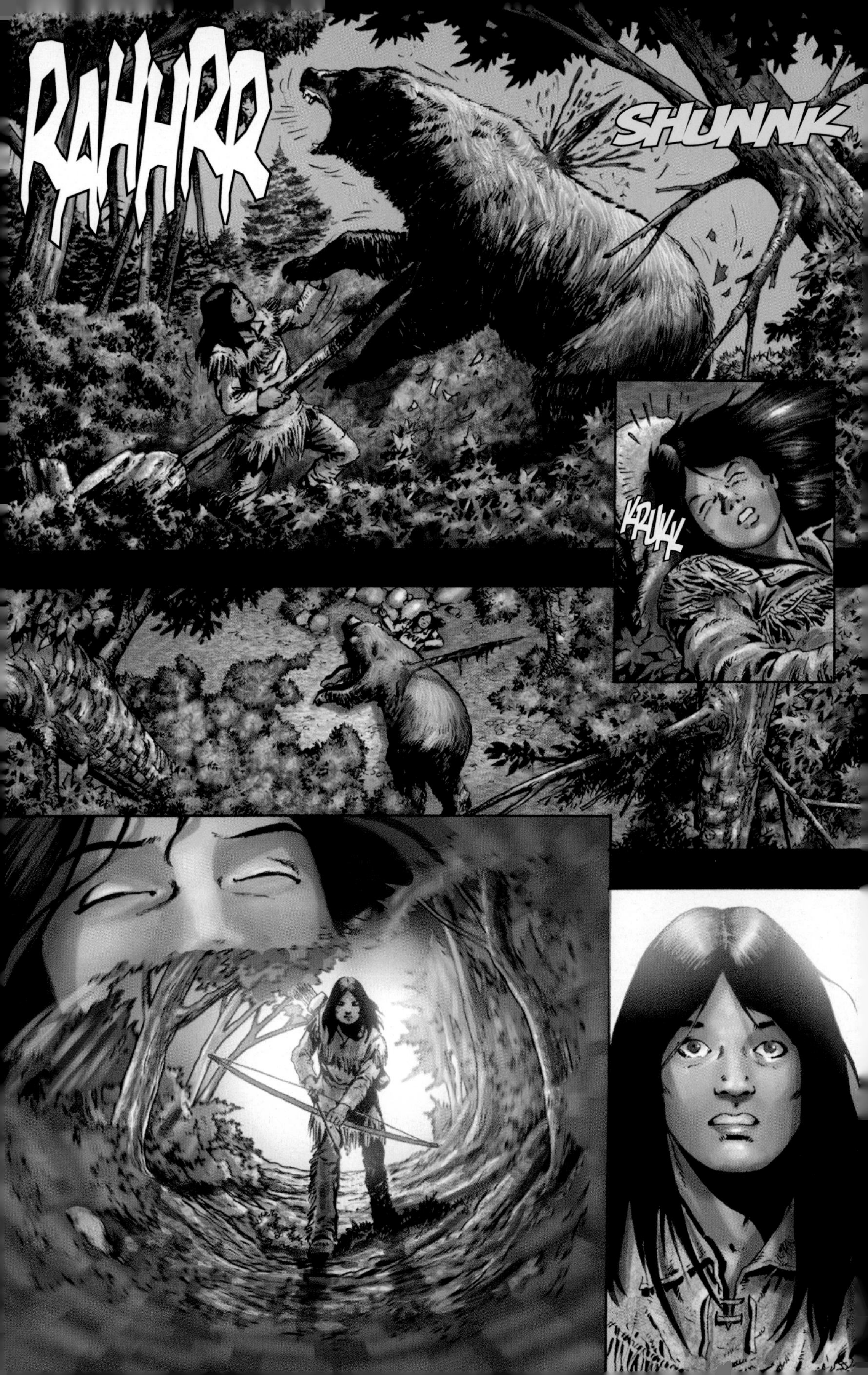
RAHHRR
SHUNNK
KRUKK

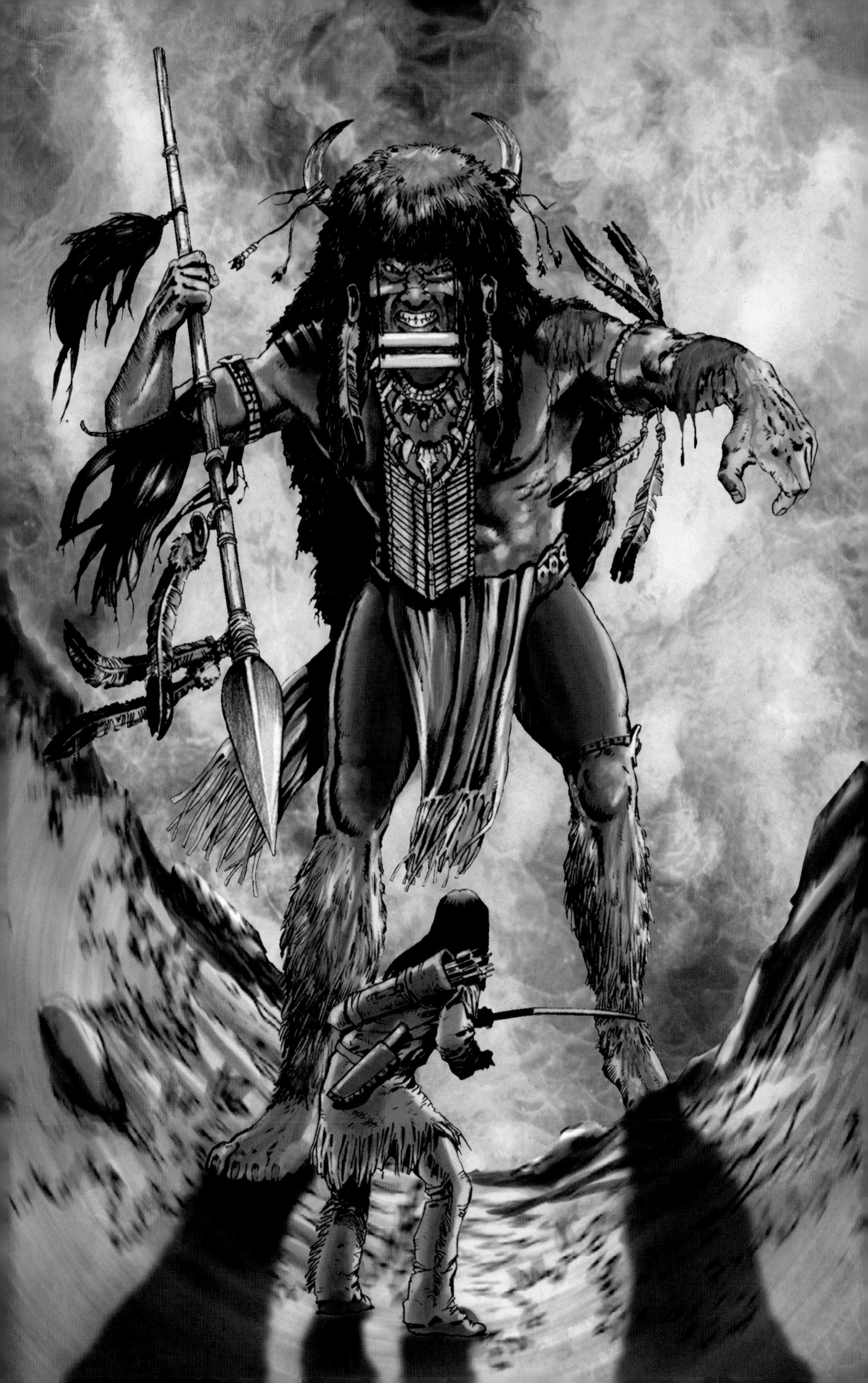

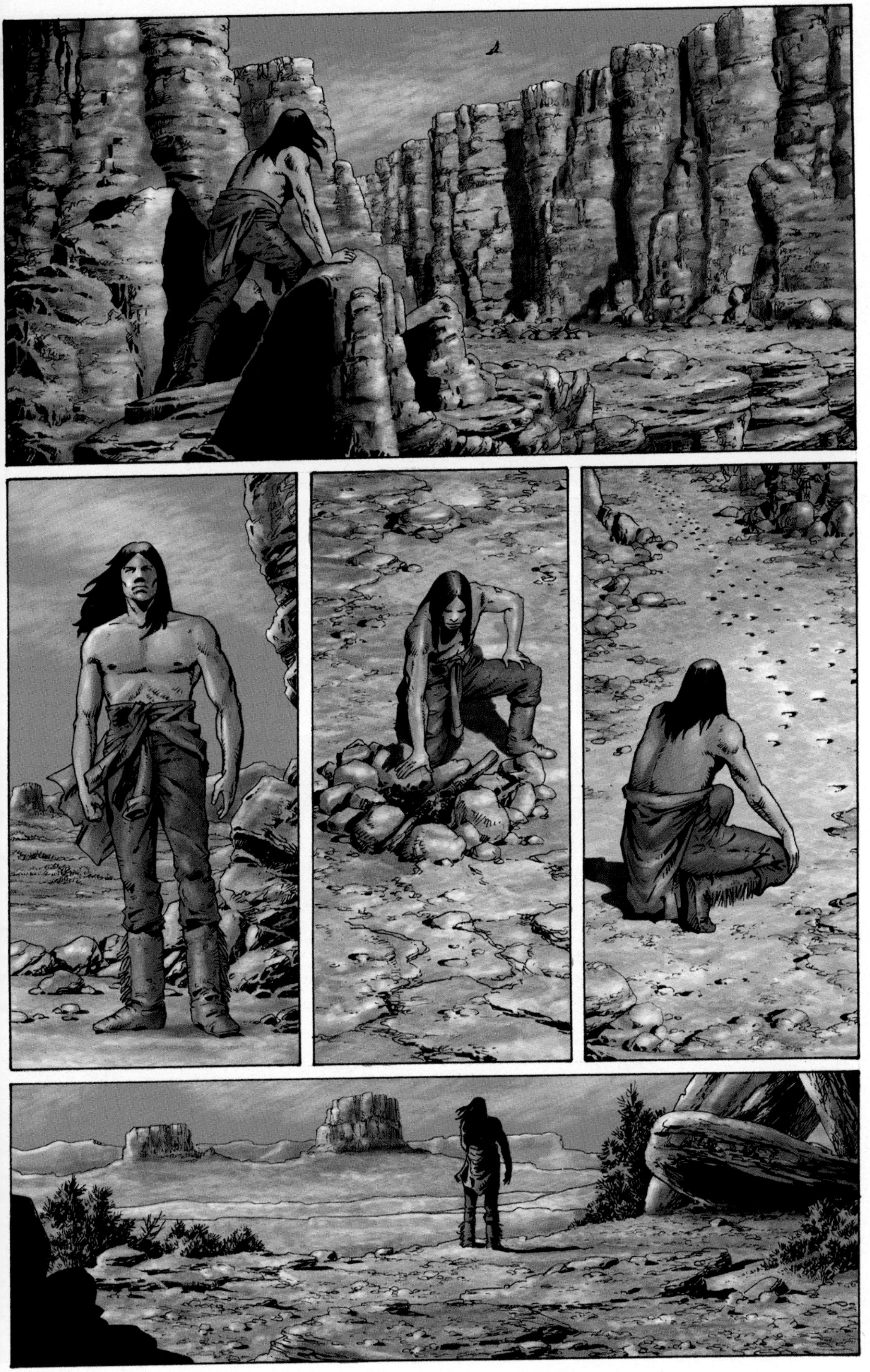

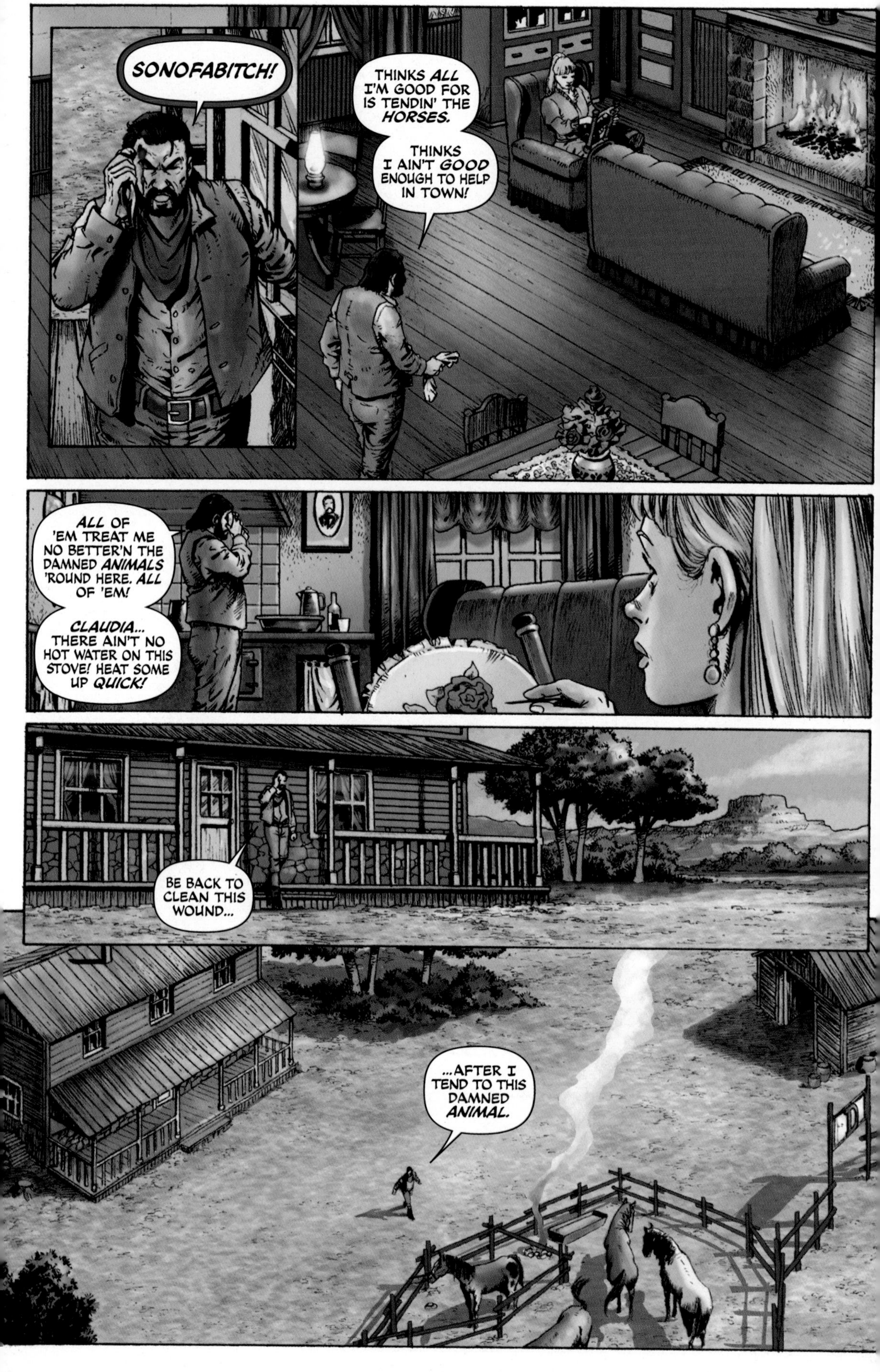
SONOFABITCH!
THINKS ALL I'M GOOD FOR IS TENDIN' THE HORSES.
THINKS I AIN'T GOOD ENOUGH TO HELP IN TOWN!
ALL OF 'EM TREAT ME NO BETTER'N THE DAMNED ANIMALS 'ROUND HERE. ALL OF 'EM!
CLAUDIA... THERE AIN'T NO HOT WATER ON THIS STOVE! HEAT SOME UP QUICK!
BE BACK TO CLEAN THIS WOUND...
...AFTER I TEND TO THIS DAMNED ANIMAL.

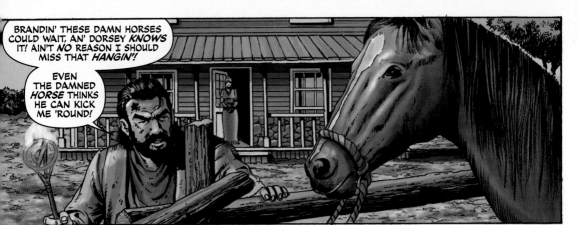

BRANDIN' THESE DAMN HORSES COULD WAIT, AN' DORSEY *KNOWS* IT! AIN'T *NO* REASON I SHOULD MISS THAT *HANGIN'!*

EVEN THE DAMNED *HORSE* THINKS HE CAN KICK ME 'ROUND!

WELL, GET READY, YOU BIG *BASTARD...*

GONNA GIVE YOU A BRAND *NO ONE* COULD MISS.

HUH...?

WHO YOU LOOKING AT OUT THERE?

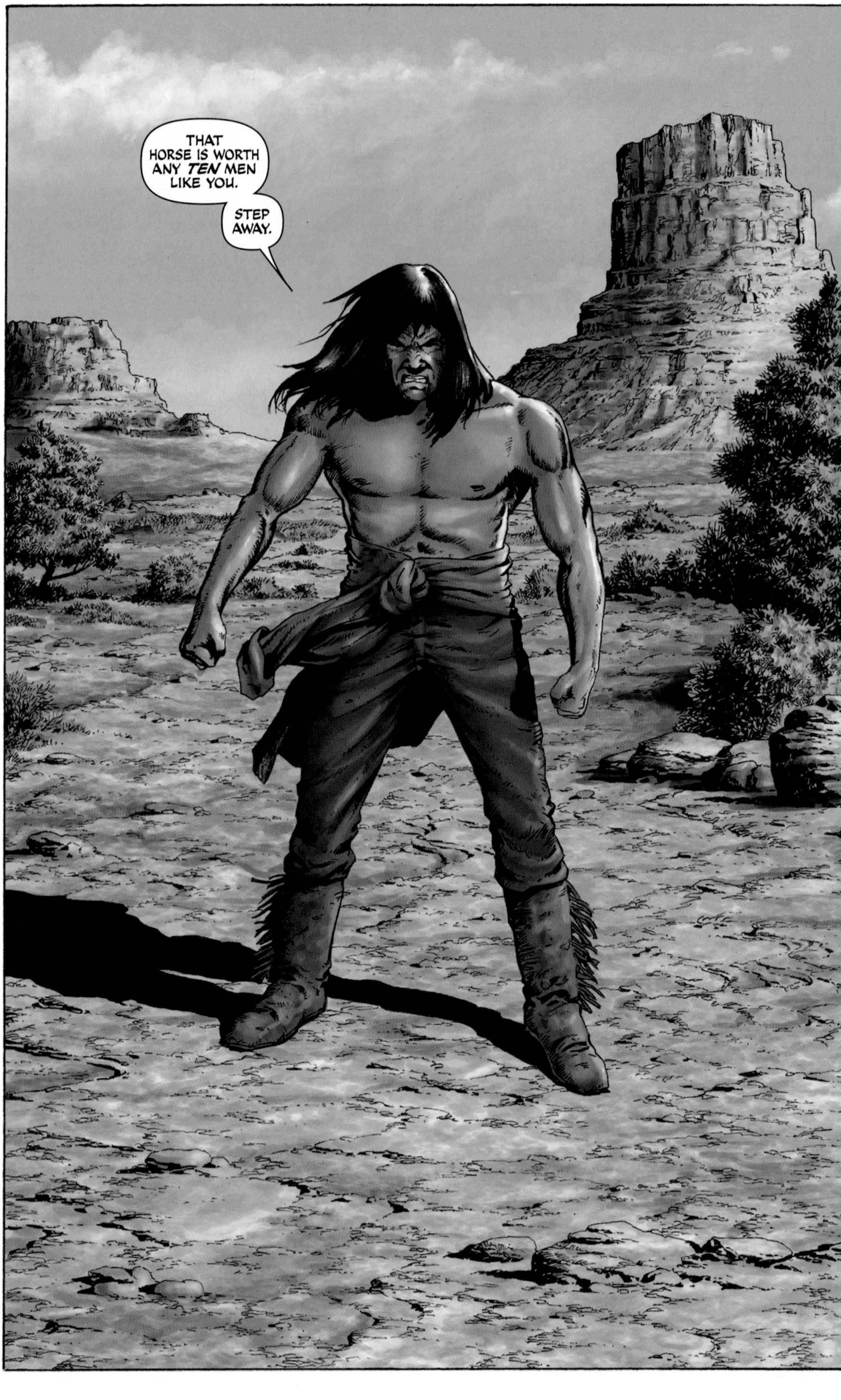
THAT HORSE IS WORTH ANY *TEN* MEN LIKE YOU.
STEP AWAY.

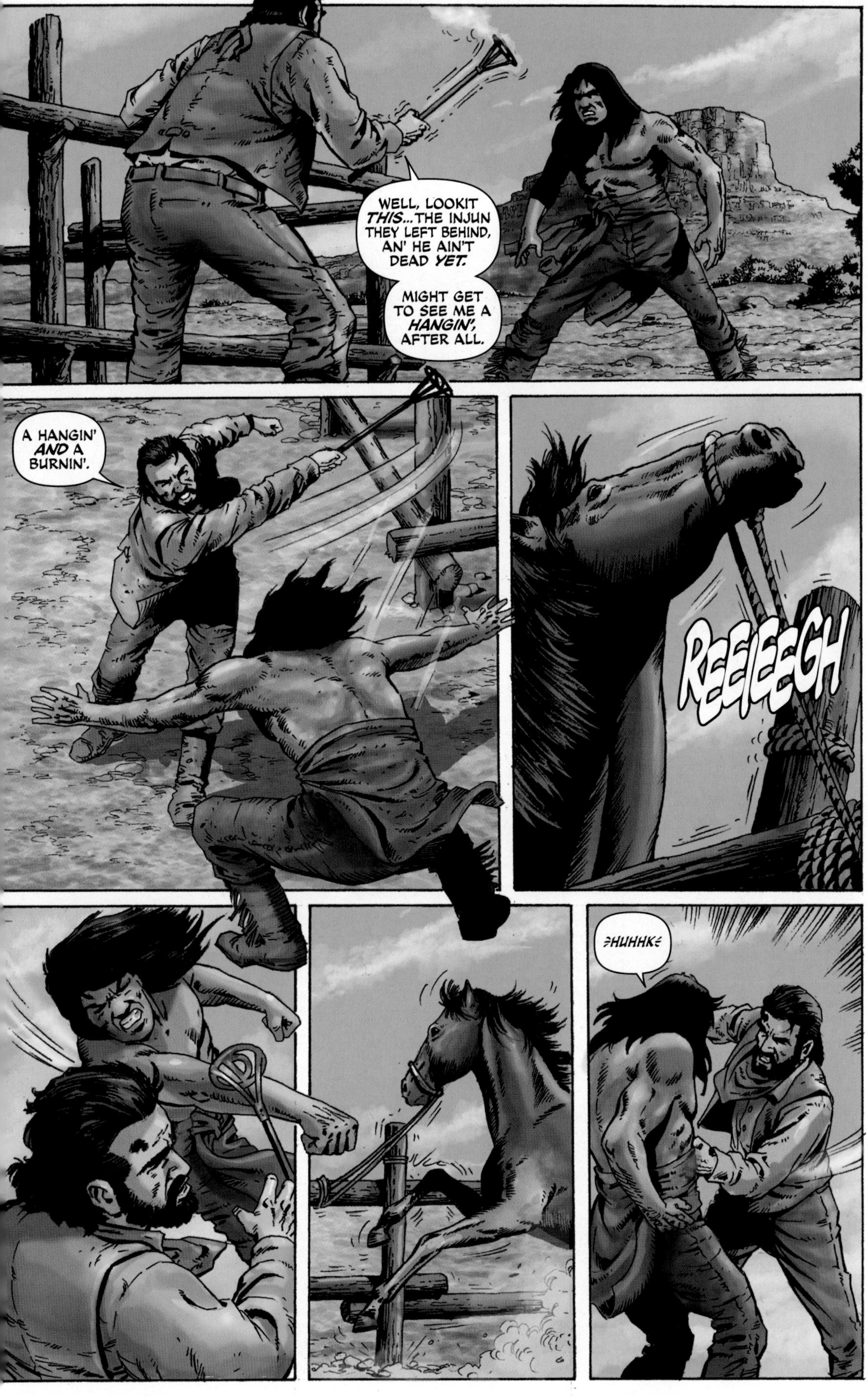
WELL, LOOKIT THIS... THE INJUN THEY LEFT BEHIND, AN' HE AIN'T DEAD YET.
MIGHT GET TO SEE ME A HANGIN', AFTER ALL.
A HANGIN' AND A BURNIN'.
REEIEEGH
HUHHK

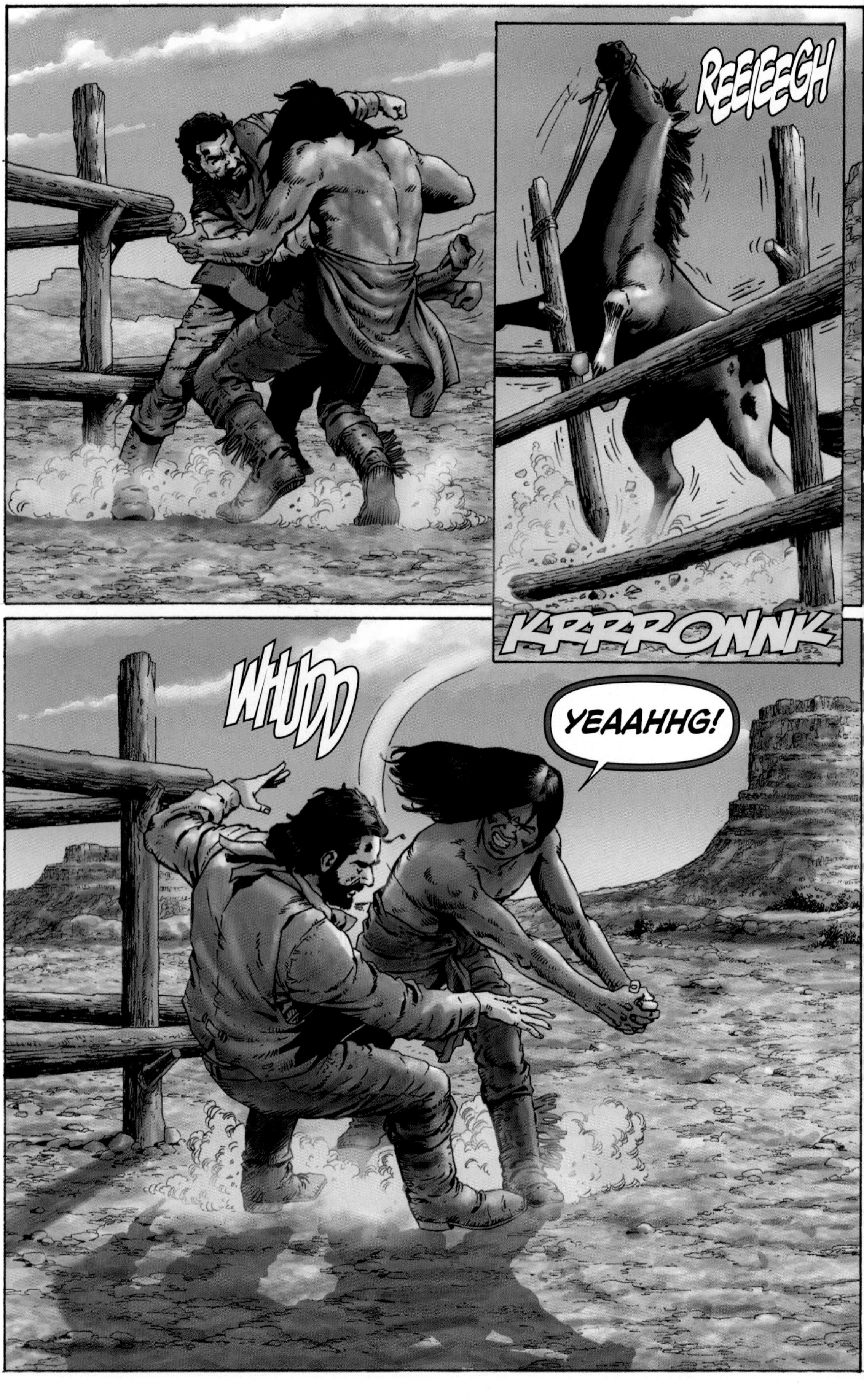
REEIEEGH
KRRRONNK
WHUDD
YEAAHHG!

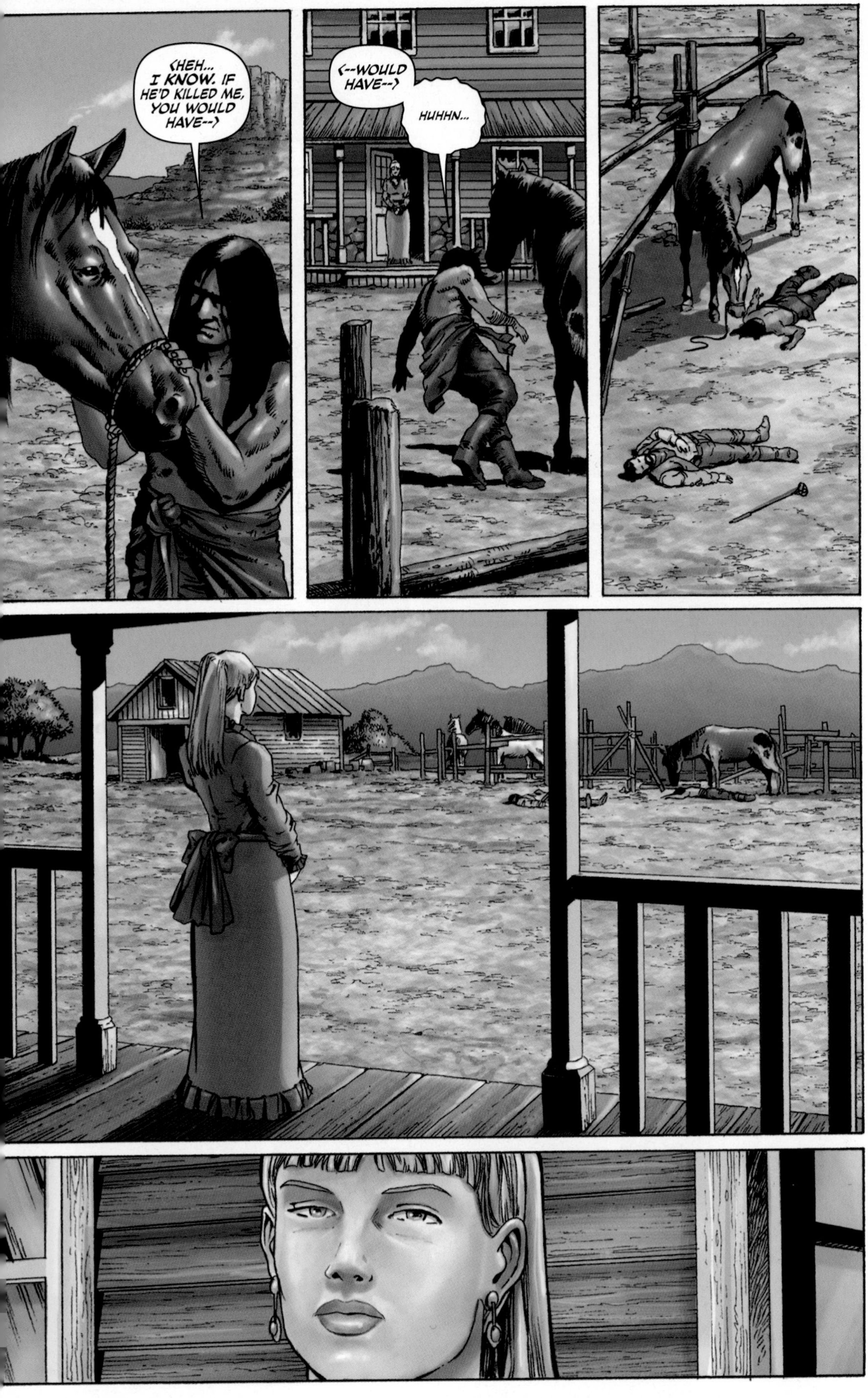
<HEH... I KNOW. IF HE'D KILLED ME, YOU WOULD HAVE-->
<--WOULD HAVE-->
HUHHN...

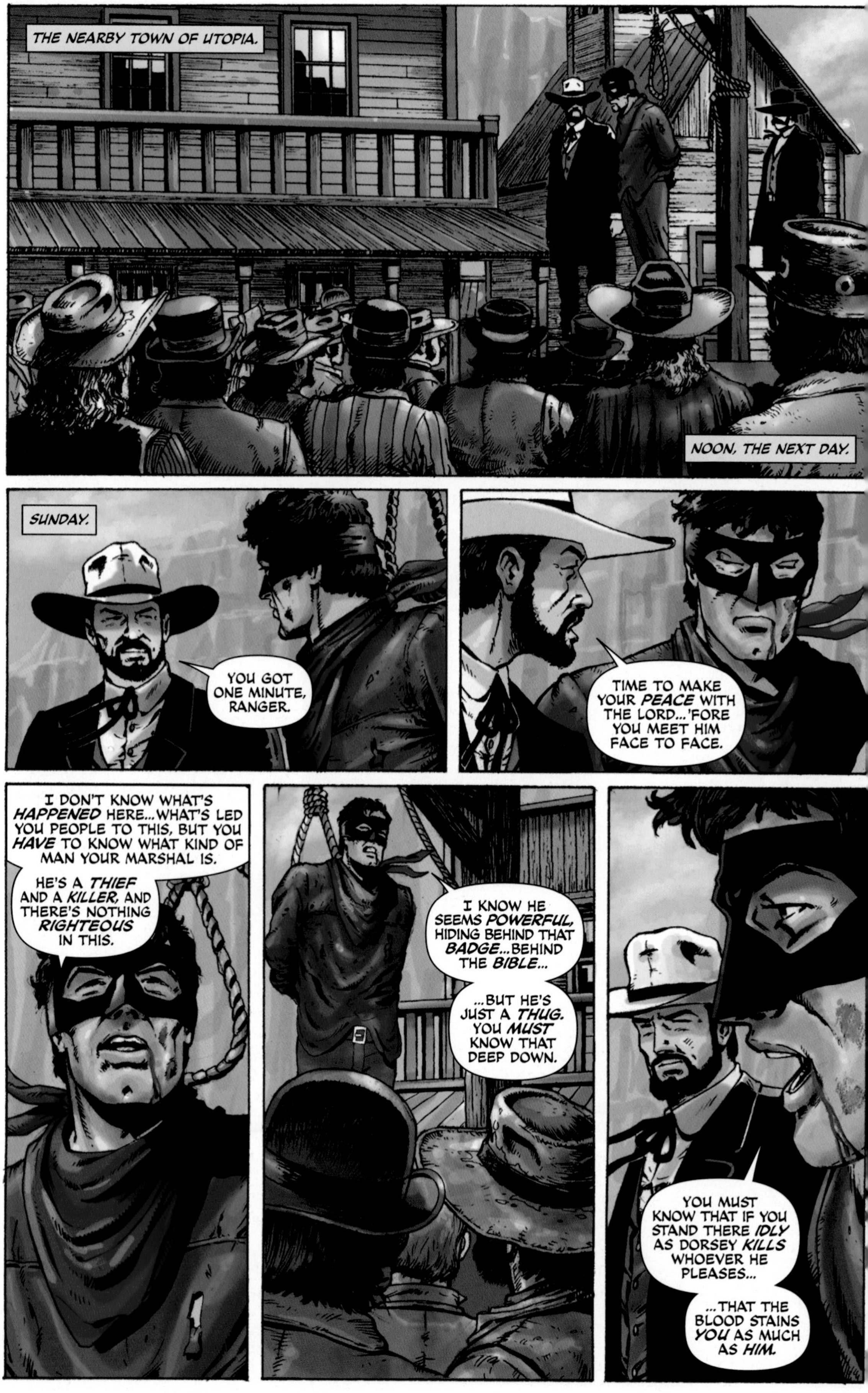
THE NEARBY TOWN OF UTOPIA.
NOON, THE NEXT DAY.
SUNDAY.
YOU GOT ONE MINUTE, RANGER.
TIME TO MAKE YOUR PEACE WITH THE LORD...'FORE YOU MEET HIM FACE TO FACE.
I DON'T KNOW WHAT'S HAPPENED HERE...WHAT'S LED YOU PEOPLE TO THIS, BUT YOU HAVE TO KNOW WHAT KIND OF MAN YOUR MARSHAL IS.
HE'S A THIEF AND A KILLER, AND THERE'S NOTHING RIGHTEOUS IN THIS.
I KNOW HE SEEMS POWERFUL, HIDING BEHIND THAT BADGE...BEHIND THE BIBLE...
...BUT HE'S JUST A THUG. YOU MUST KNOW THAT DEEP DOWN.
YOU MUST KNOW THAT IF YOU STAND THERE IDLY AS DORSEY KILLS WHOEVER HE PLEASES...
...THAT THE BLOOD STAINS YOU AS MUCH AS HIM.

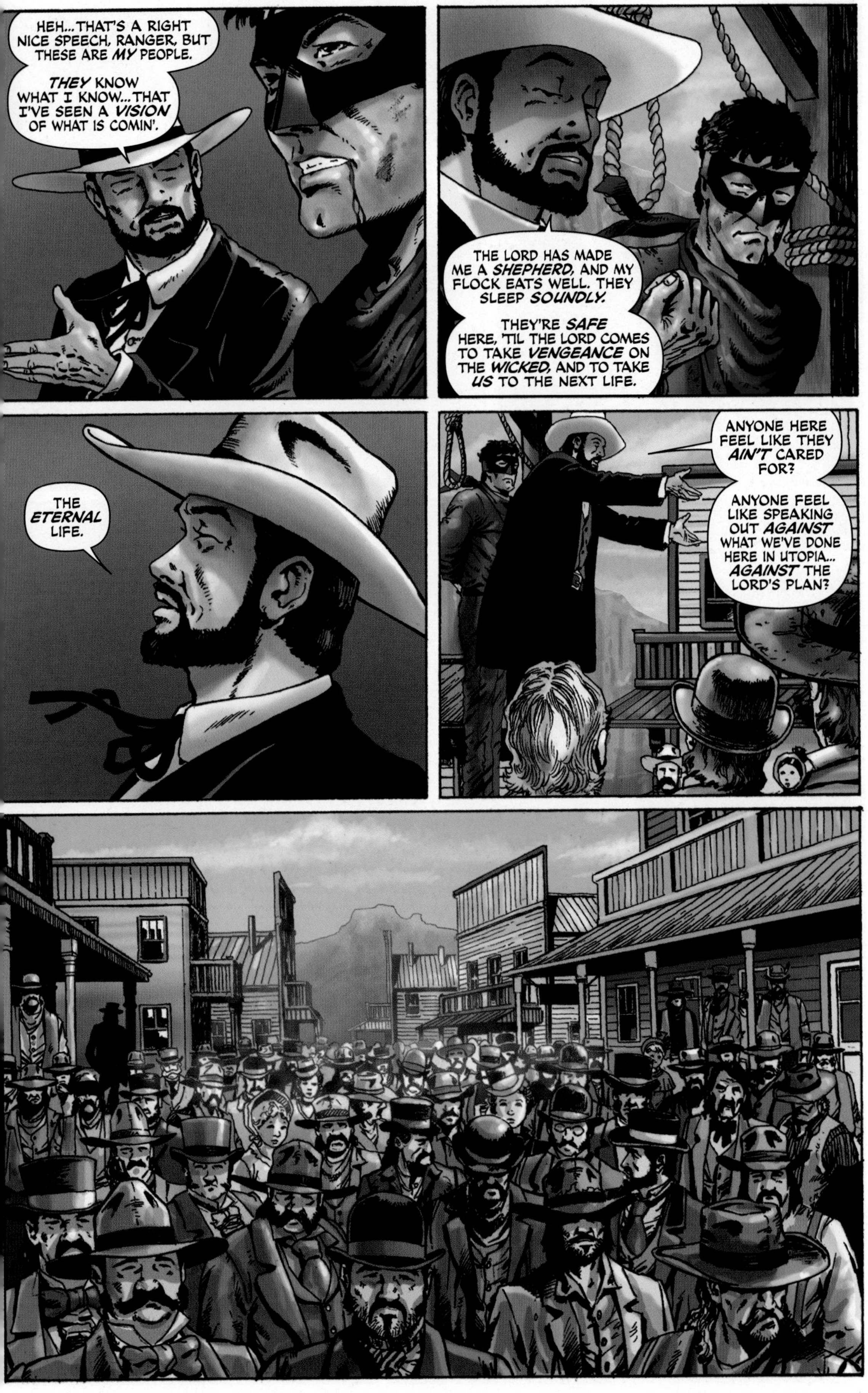
HEH...THAT'S A RIGHT NICE SPEECH, RANGER, BUT THESE ARE *MY* PEOPLE.
THEY KNOW WHAT I KNOW...THAT I'VE SEEN A *VISION* OF WHAT IS COMIN'.
THE LORD HAS MADE ME A *SHEPHERD,* AND MY FLOCK EATS WELL. THEY SLEEP *SOUNDLY.*
THEY'RE *SAFE* HERE, 'TIL THE LORD COMES TO TAKE *VENGEANCE* ON THE *WICKED,* AND TO TAKE *US* TO THE NEXT LIFE.
THE *ETERNAL* LIFE.
ANYONE HERE FEEL LIKE THEY *AIN'T* CARED FOR?
ANYONE FEEL LIKE SPEAKING OUT *AGAINST* WHAT WE'VE DONE HERE IN UTOPIA... *AGAINST* THE LORD'S PLAN?

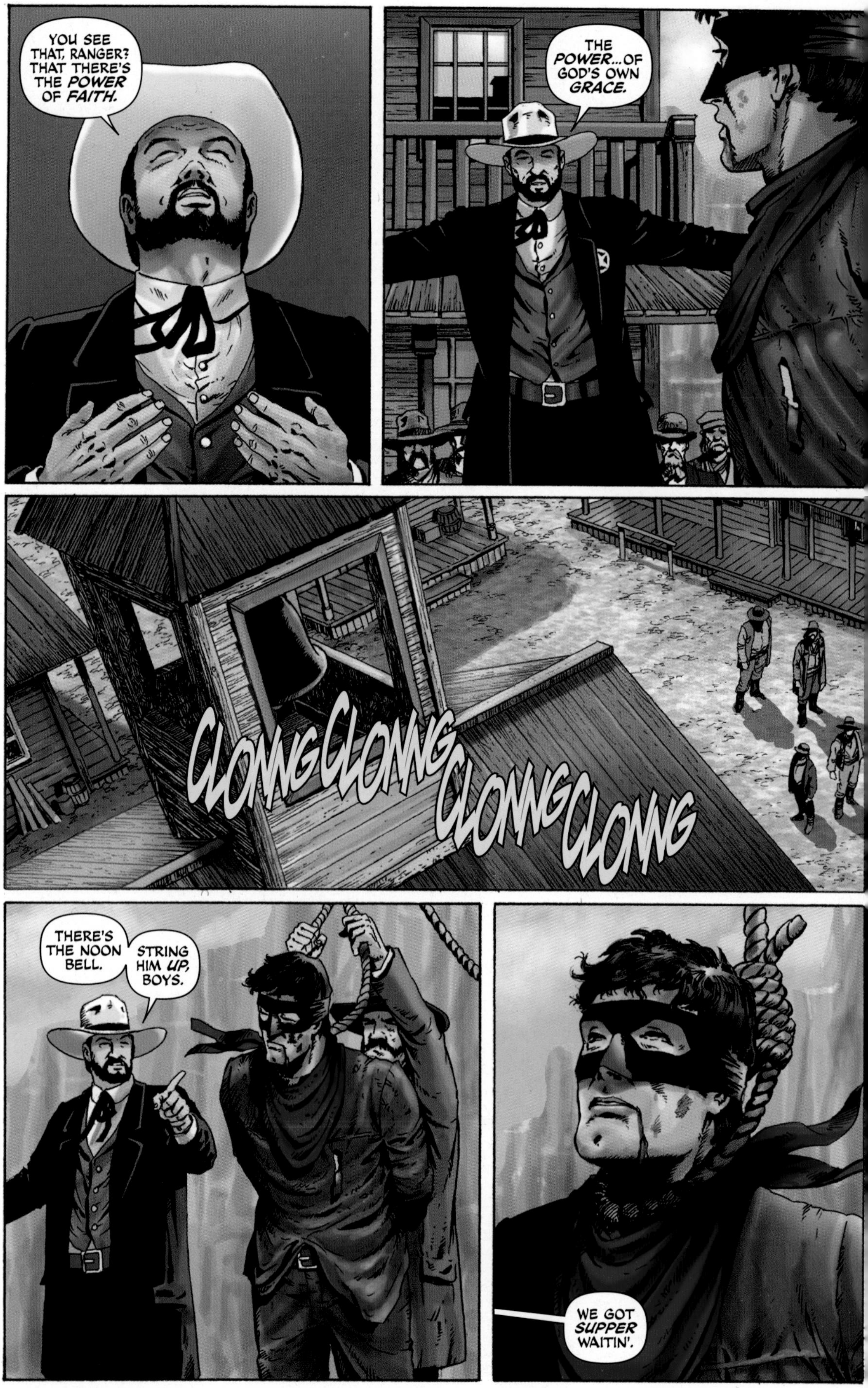
YOU SEE THAT, RANGER? THAT THERE'S THE *POWER* OF *FAITH.*
THE *POWER*...OF GOD'S OWN *GRACE.*
CLONNG CLONNG CLONNG CLONNG
THERE'S THE NOON BELL.
STRING HIM *UP,* BOYS.
WE GOT *SUPPER* WAITIN'.

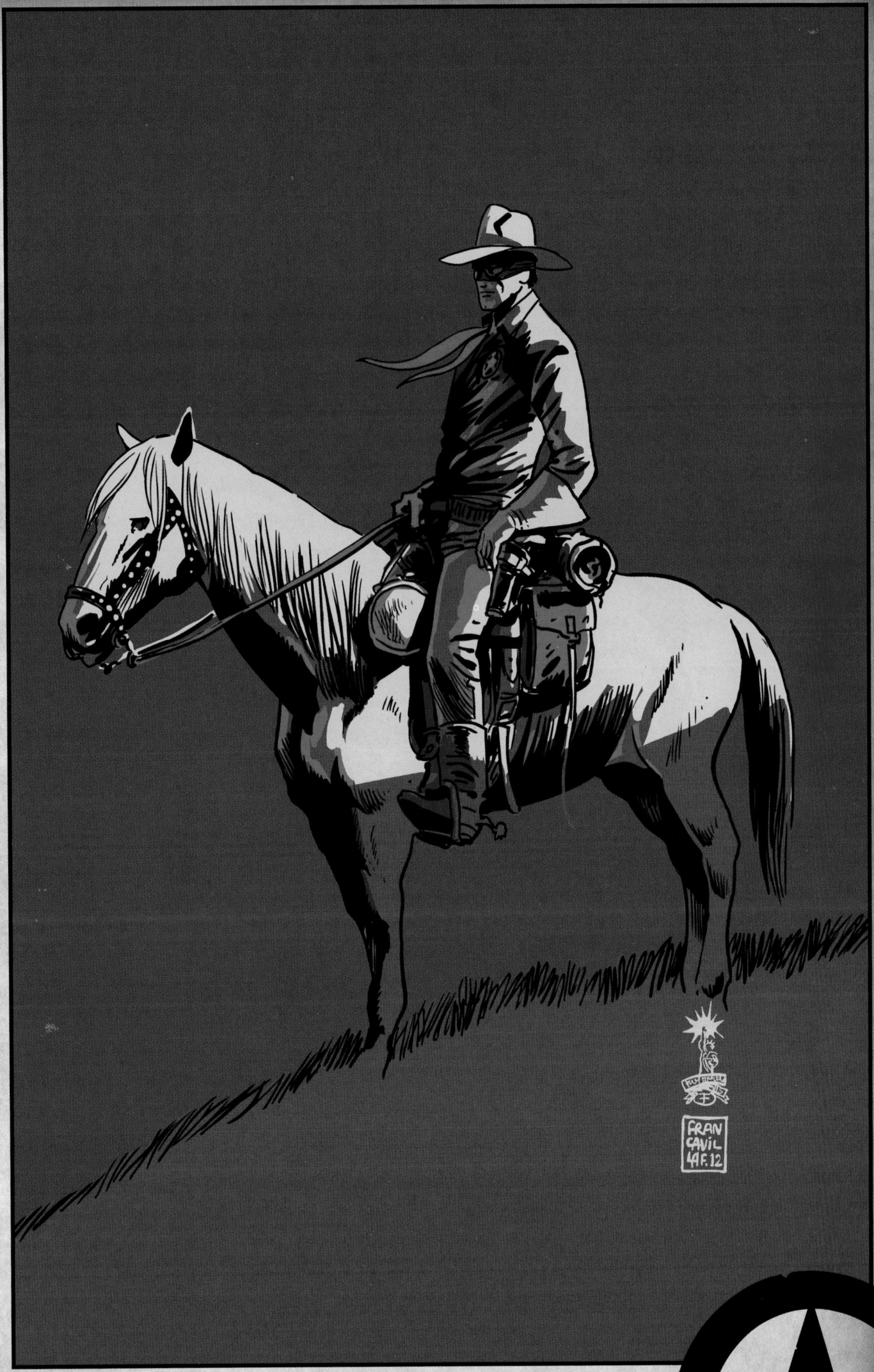
FRAN
CAVIL
LA F. 12

COVER BY **FRANCESCO FRANCAVILLA**

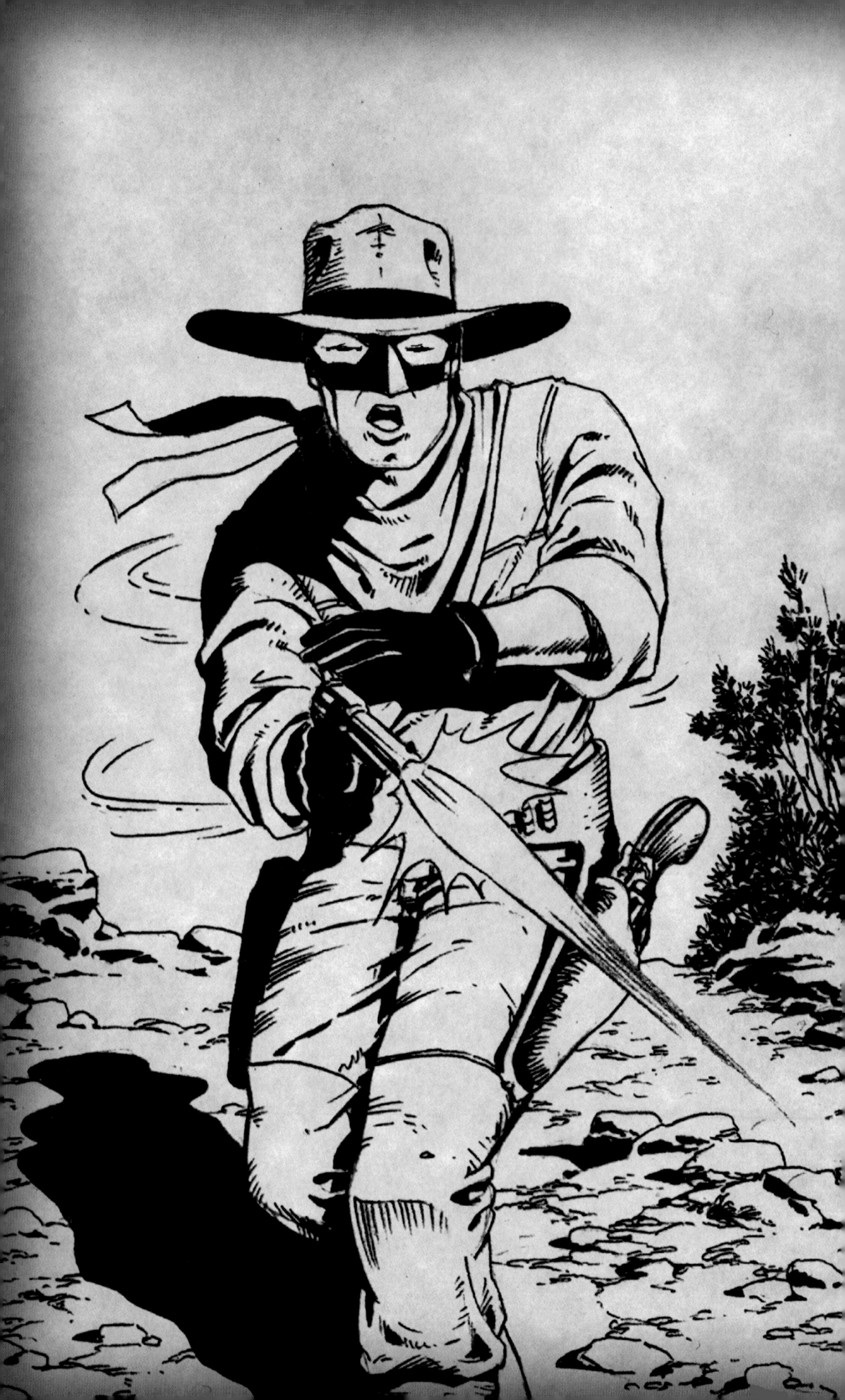

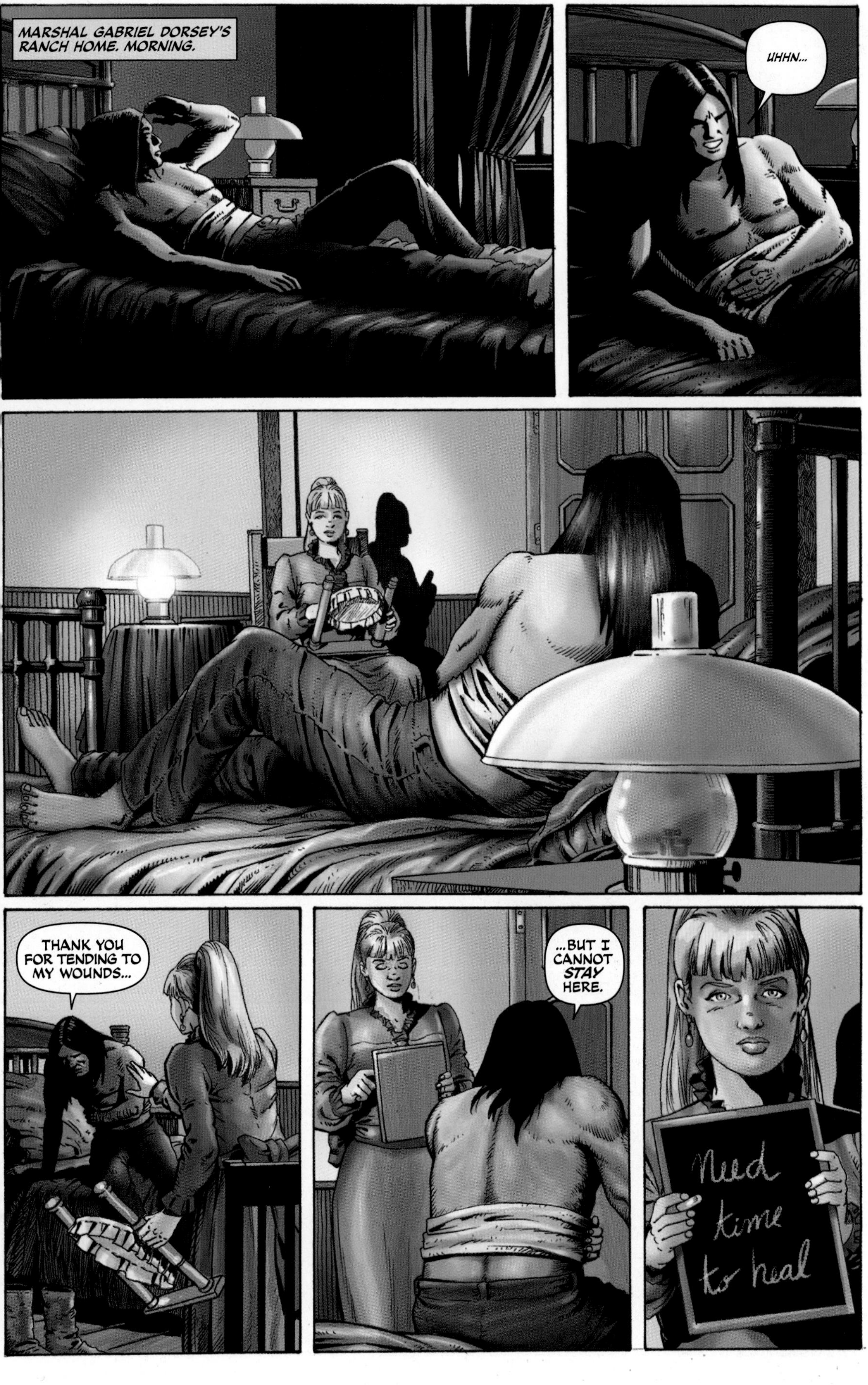
MARSHAL GABRIEL DORSEY'S RANCH HOME. MORNING.
UHHN...
THANK YOU FOR TENDING TO MY WOUNDS...
...BUT I CANNOT STAY HERE.
Need time to heal

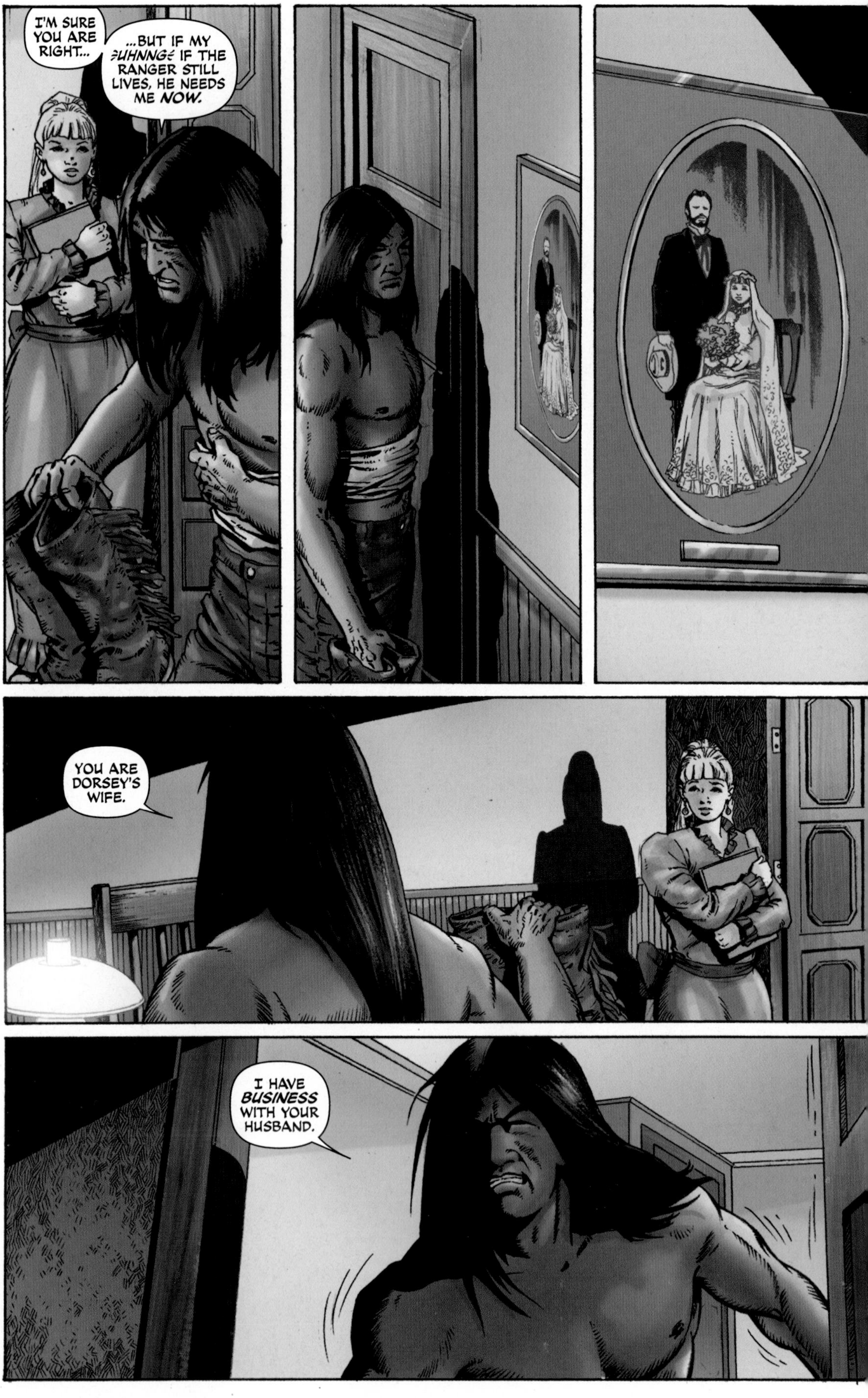

I'M SURE YOU ARE RIGHT...
...BUT IF MY =UHNNG= IF THE RANGER STILL LIVES, HE NEEDS ME *NOW*.
YOU ARE DORSEY'S WIFE.
I HAVE *BUSINESS* WITH YOUR HUSBAND.

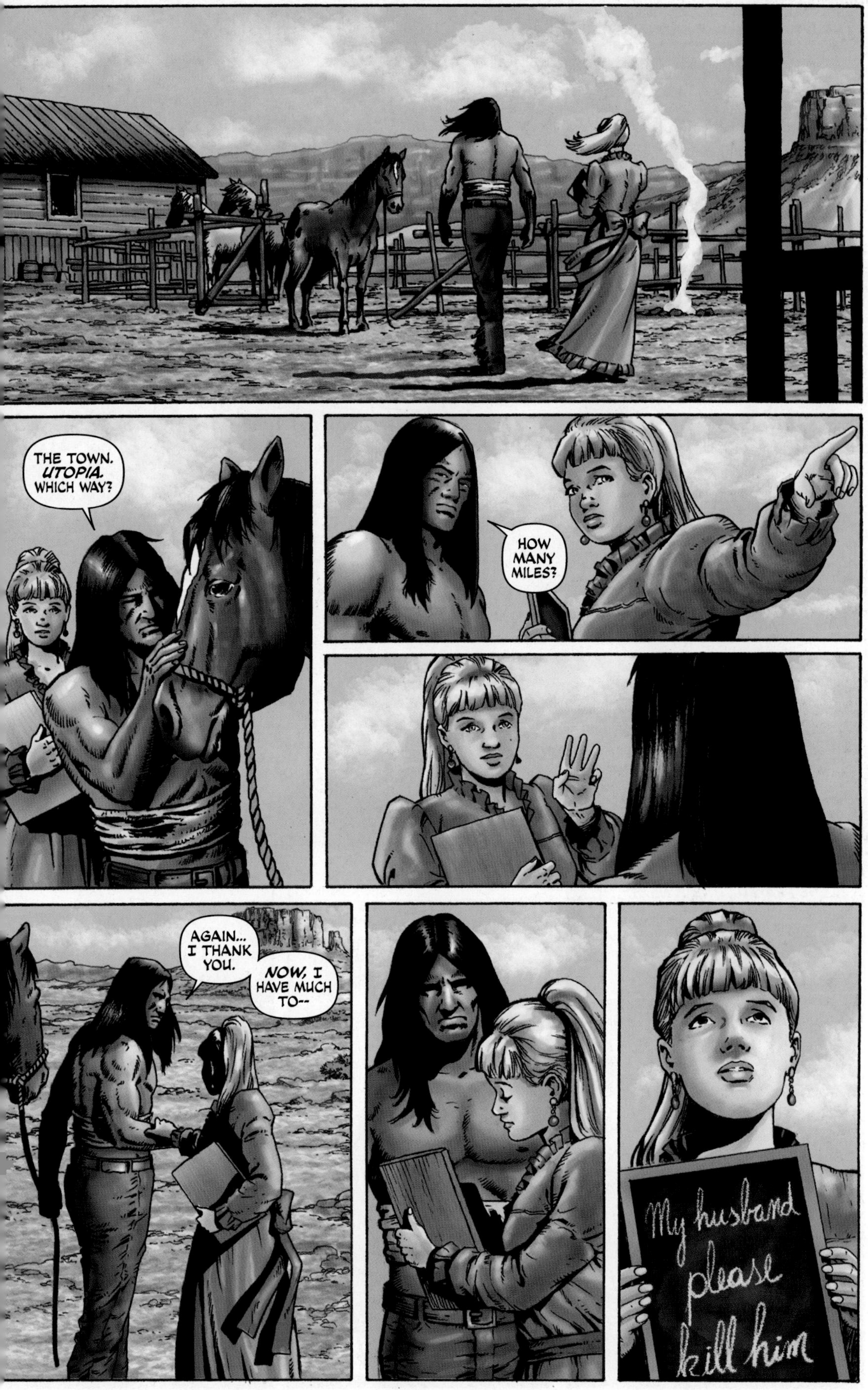
THE TOWN. *UTOPIA.* WHICH WAY?
HOW MANY MILES?
AGAIN... I THANK YOU.
NOW, I HAVE MUCH TO--
My husband please kill him

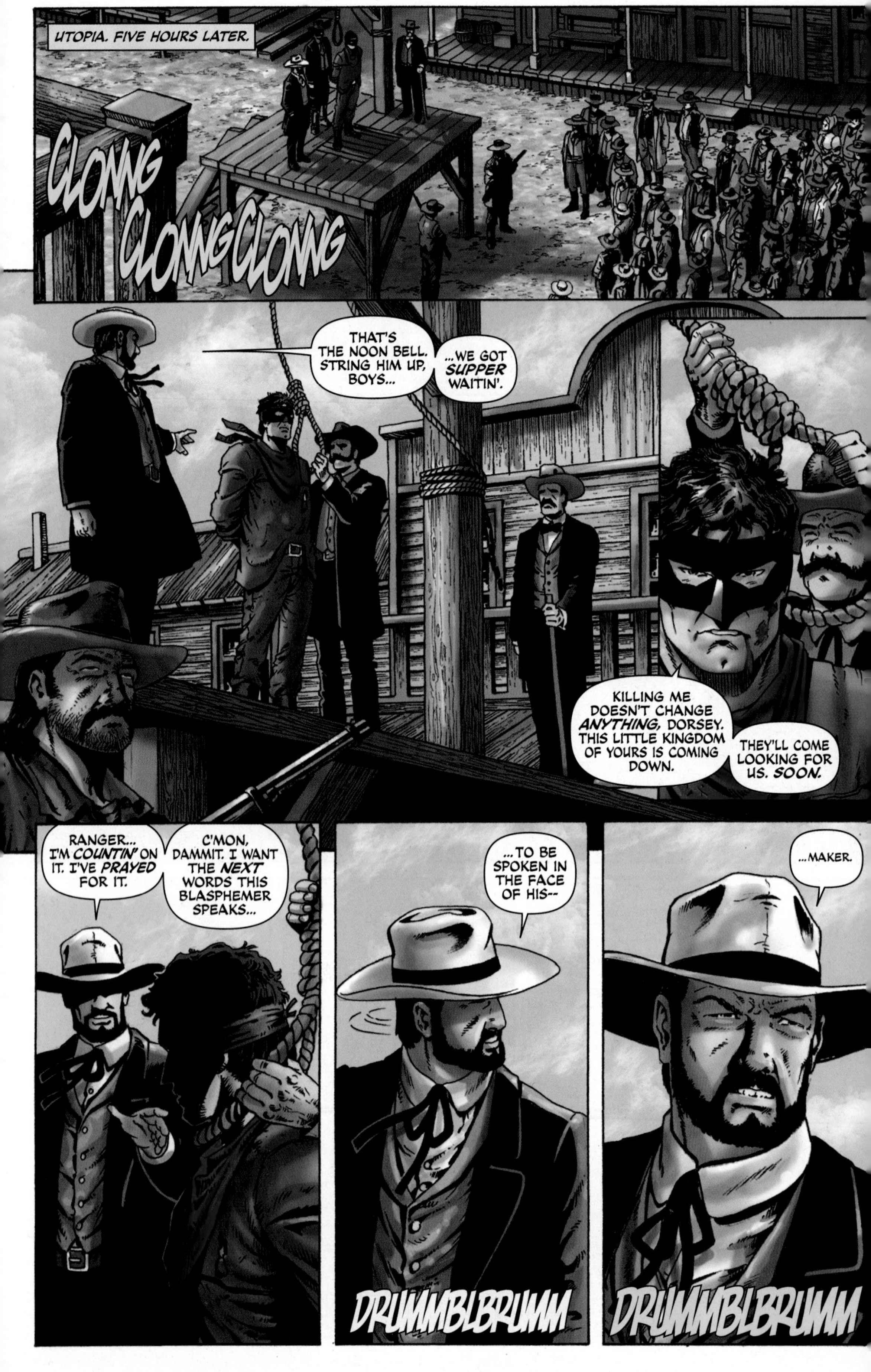
UTOPIA. FIVE HOURS LATER.
CLONNG
CLONNG CLONNG CLONNG
THAT'S THE NOON BELL. STRING HIM UP, BOYS...
...WE GOT SUPPER WAITIN'.
KILLING ME DOESN'T CHANGE ANYTHING, DORSEY. THIS LITTLE KINGDOM OF YOURS IS COMING DOWN.
THEY'LL COME LOOKING FOR US. SOON.
RANGER... I'M COUNTIN' ON IT. I'VE PRAYED FOR IT.
C'MON, DAMMIT. I WANT THE NEXT WORDS THIS BLASPHEMER SPEAKS...
...TO BE SPOKEN IN THE FACE OF HIS--
DRUMMBLBRUMM
...MAKER.
DRUMMBLBRUMM

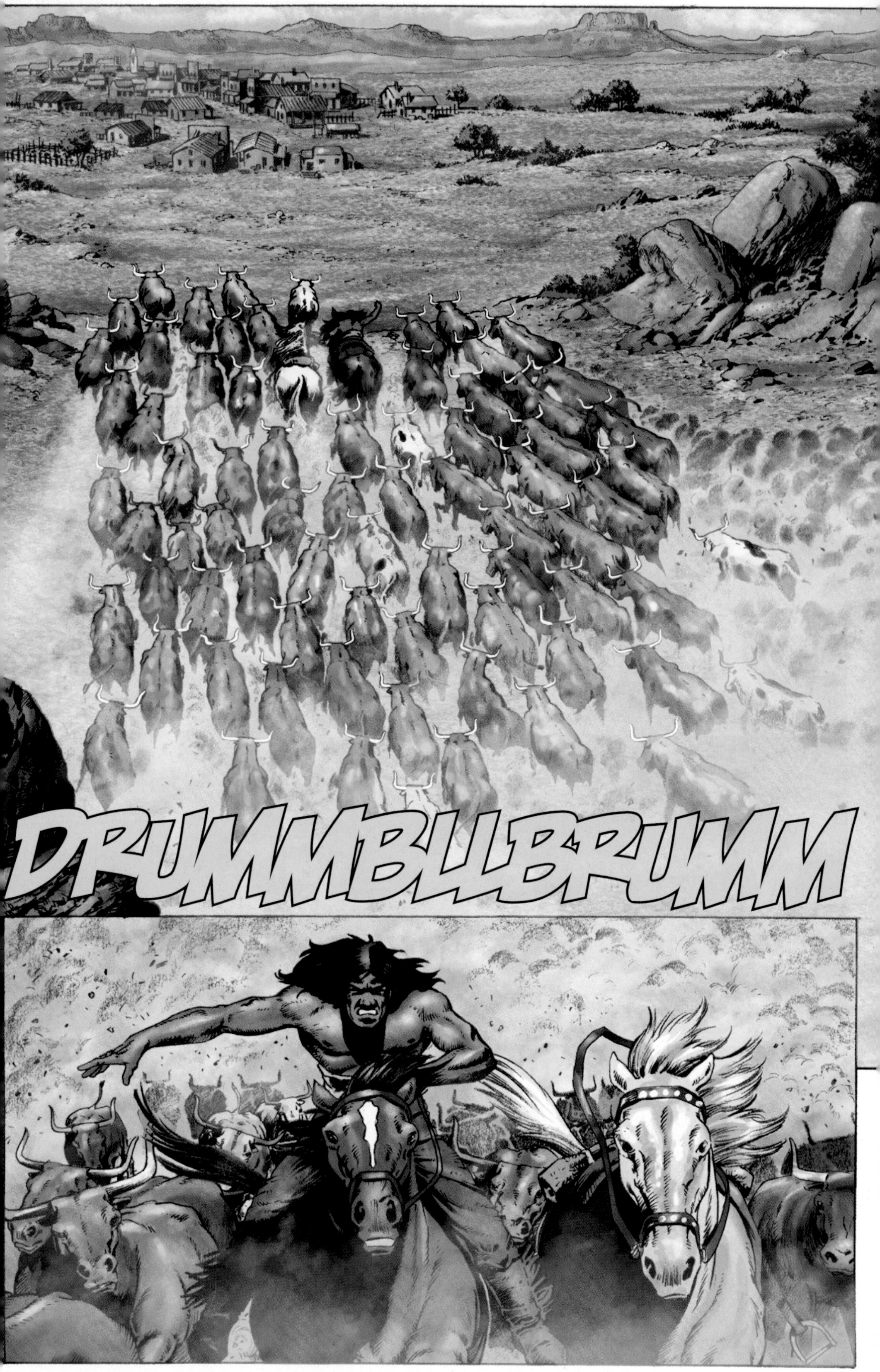
DRUMMBUUBRUMM

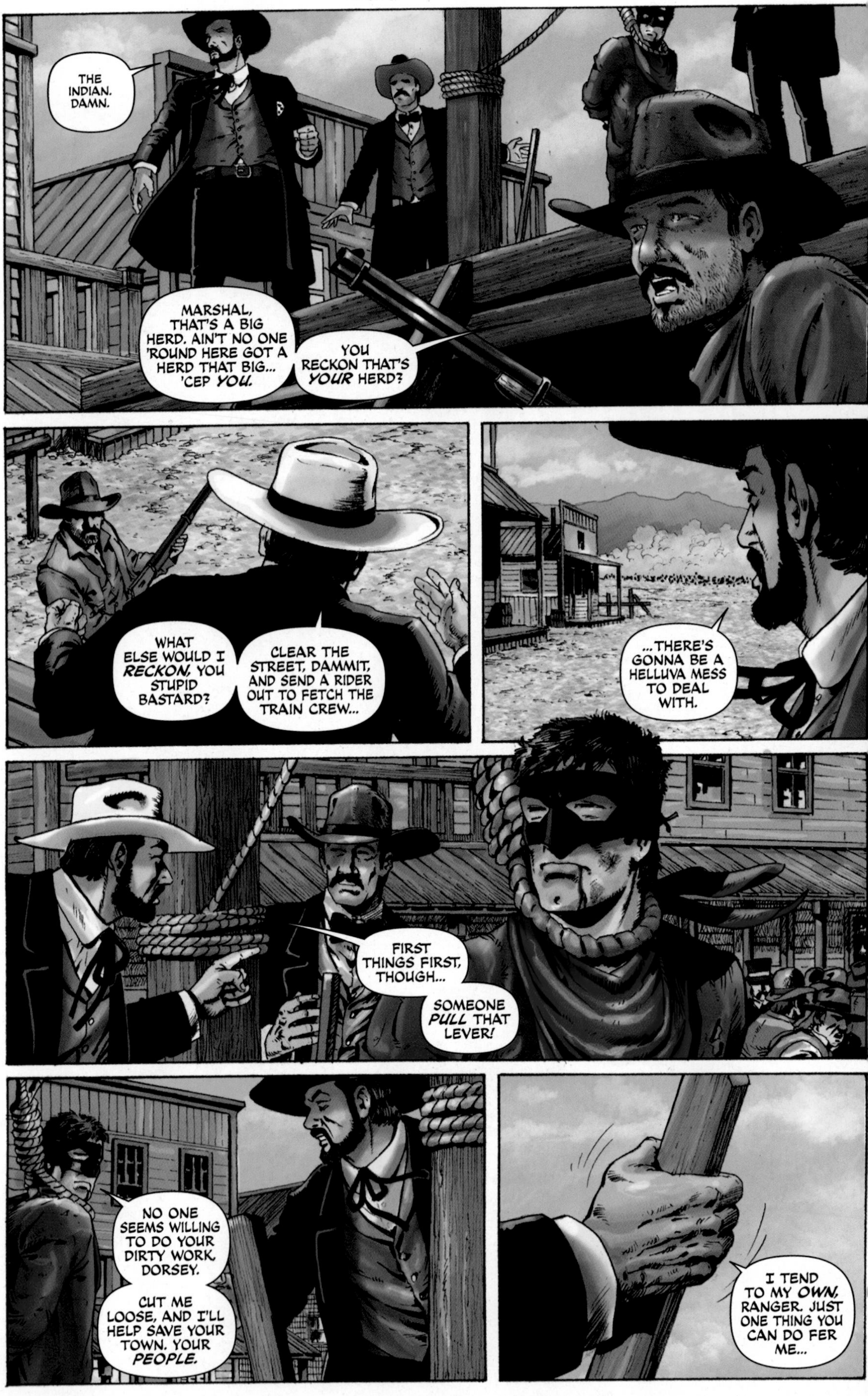
THE INDIAN. DAMN.
MARSHAL, THAT'S A BIG HERD. AIN'T NO ONE 'ROUND HERE GOT A HERD THAT BIG... 'CEP *YOU.*
YOU RECKON THAT'S *YOUR* HERD?
WHAT ELSE WOULD I *RECKON,* YOU STUPID BASTARD?
CLEAR THE STREET, DAMMIT, AND SEND A RIDER OUT TO FETCH THE TRAIN CREW...
...THERE'S GONNA BE A HELLUVA MESS TO DEAL WITH.
FIRST THINGS FIRST, THOUGH...
SOMEONE *PULL* THAT LEVER!
NO ONE SEEMS WILLING TO DO YOUR DIRTY WORK, DORSEY.
CUT ME LOOSE, AND I'LL HELP SAVE YOUR TOWN. YOUR *PEOPLE.*
I TEND TO MY *OWN,* RANGER. JUST ONE THING YOU CAN DO FER ME...

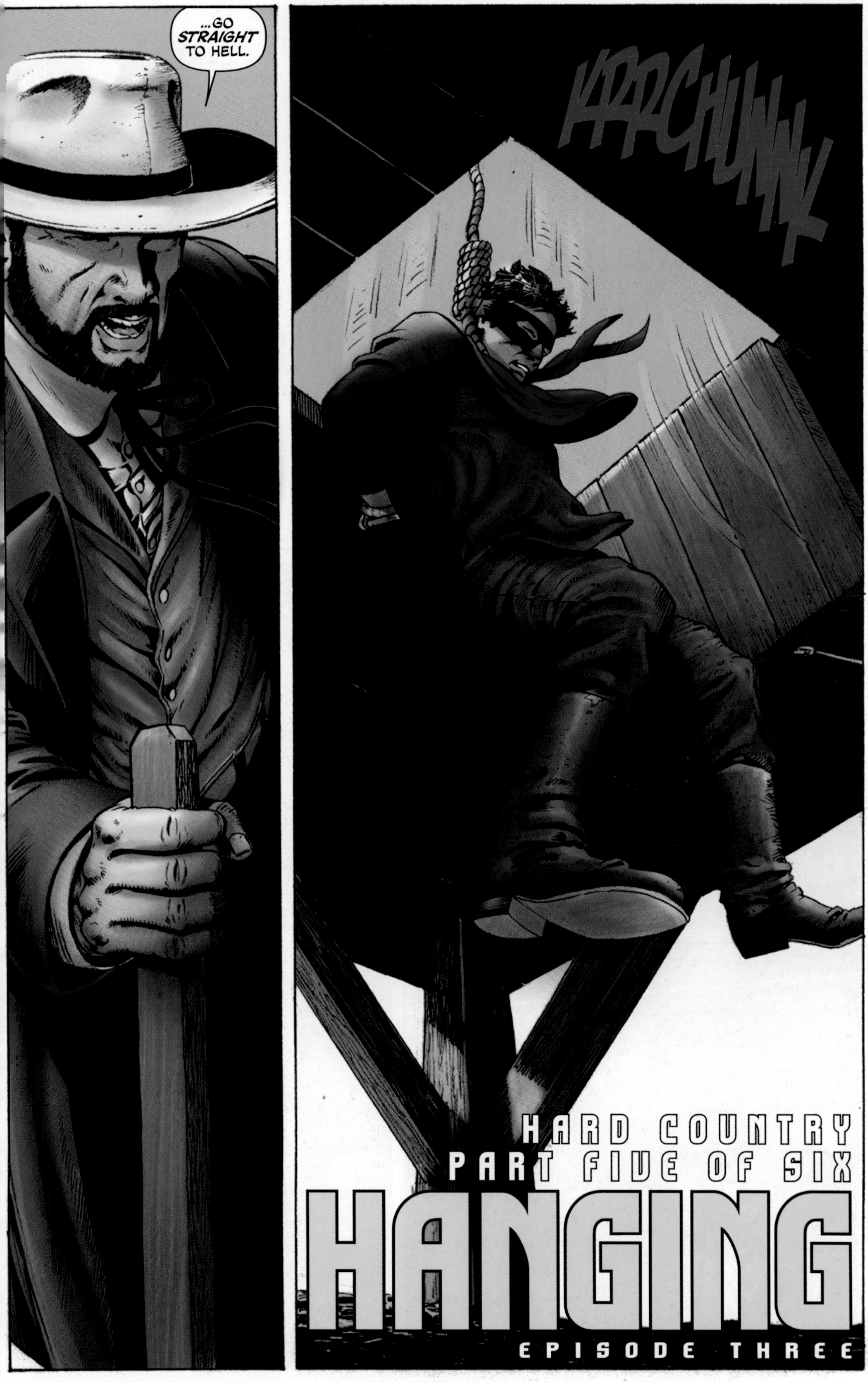
...GO STRAIGHT TO HELL.
KRRCHUNNK
HARD COUNTRY
PART FIVE OF SIX
HANGING
EPISODE THREE

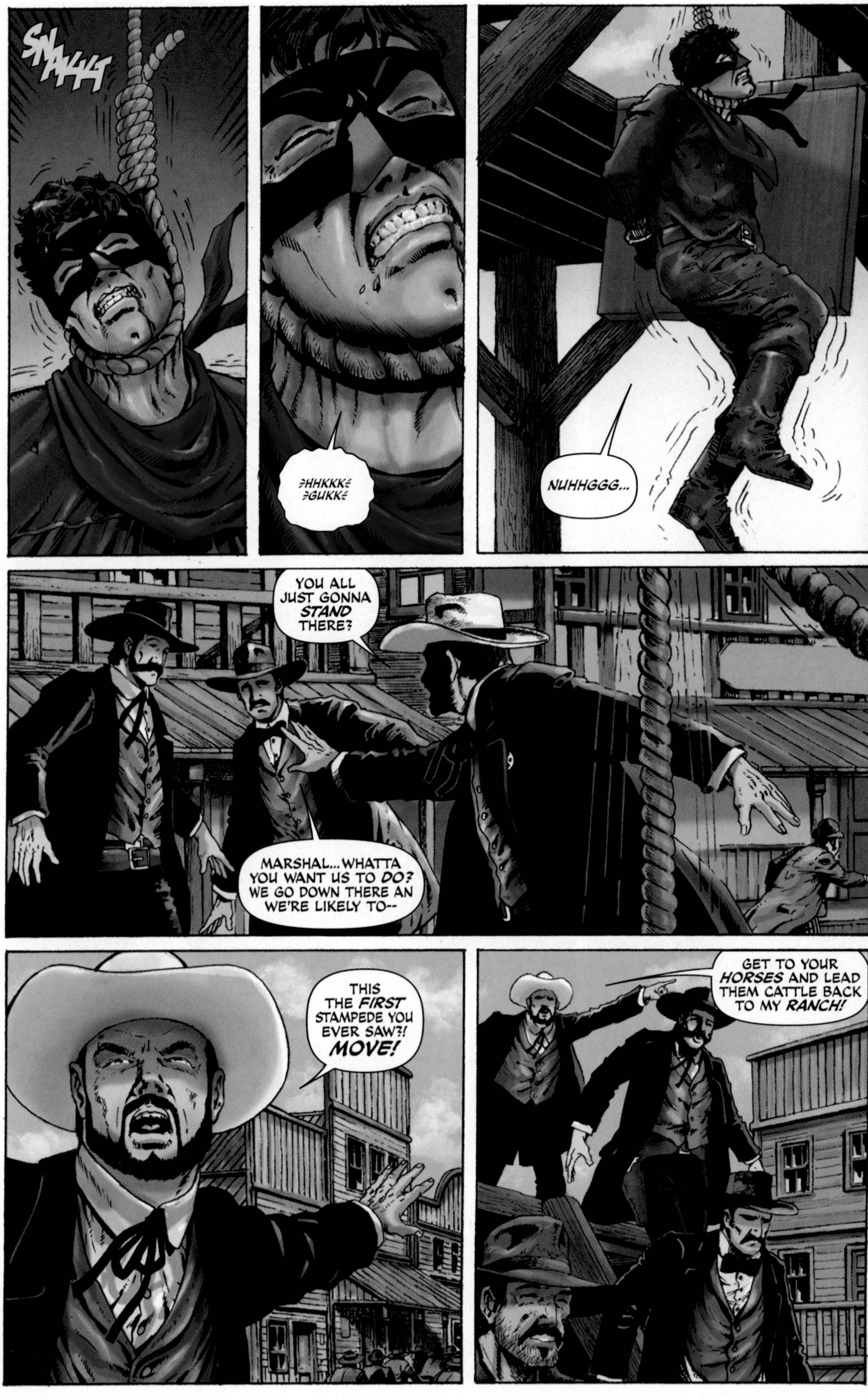
SNAKKT
HHKKK GUKK
NUHHGGG...
YOU ALL JUST GONNA STAND THERE?
MARSHAL...WHATTA YOU WANT US TO DO? WE GO DOWN THERE AN WE'RE LIKELY TO--
THIS THE FIRST STAMPEDE YOU EVER SAW?! MOVE!
GET TO YOUR HORSES AND LEAD THEM CATTLE BACK TO MY RANCH!

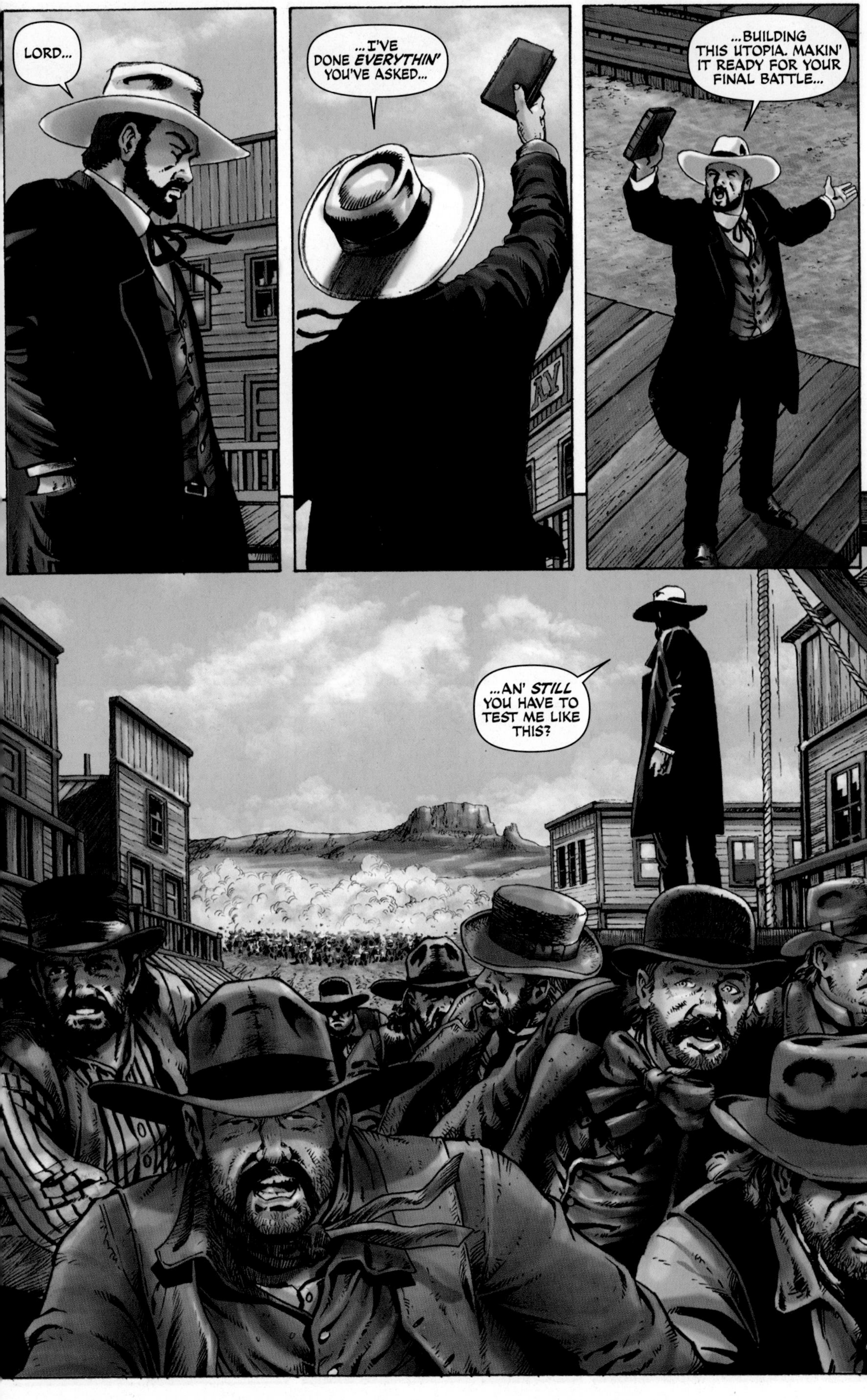
LORD...
...I'VE DONE EVERYTHIN' YOU'VE ASKED...
...BUILDING THIS UTOPIA. MAKIN' IT READY FOR YOUR FINAL BATTLE...
...AN' STILL YOU HAVE TO TEST ME LIKE THIS?

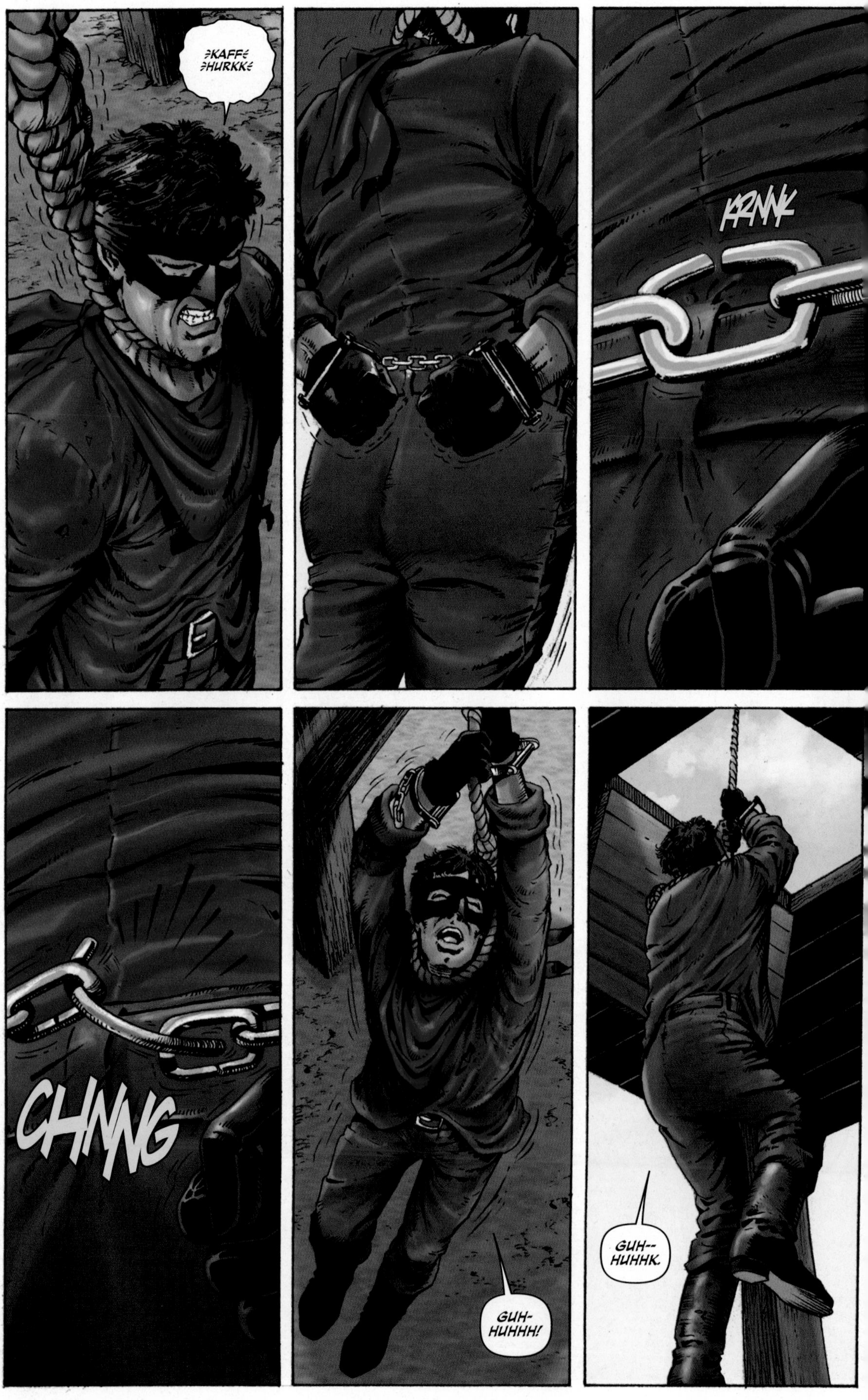
KAFF
HURKK
KRNNK
CHNNG
GUH-
HUHHH!
GUH--
HUHHK.

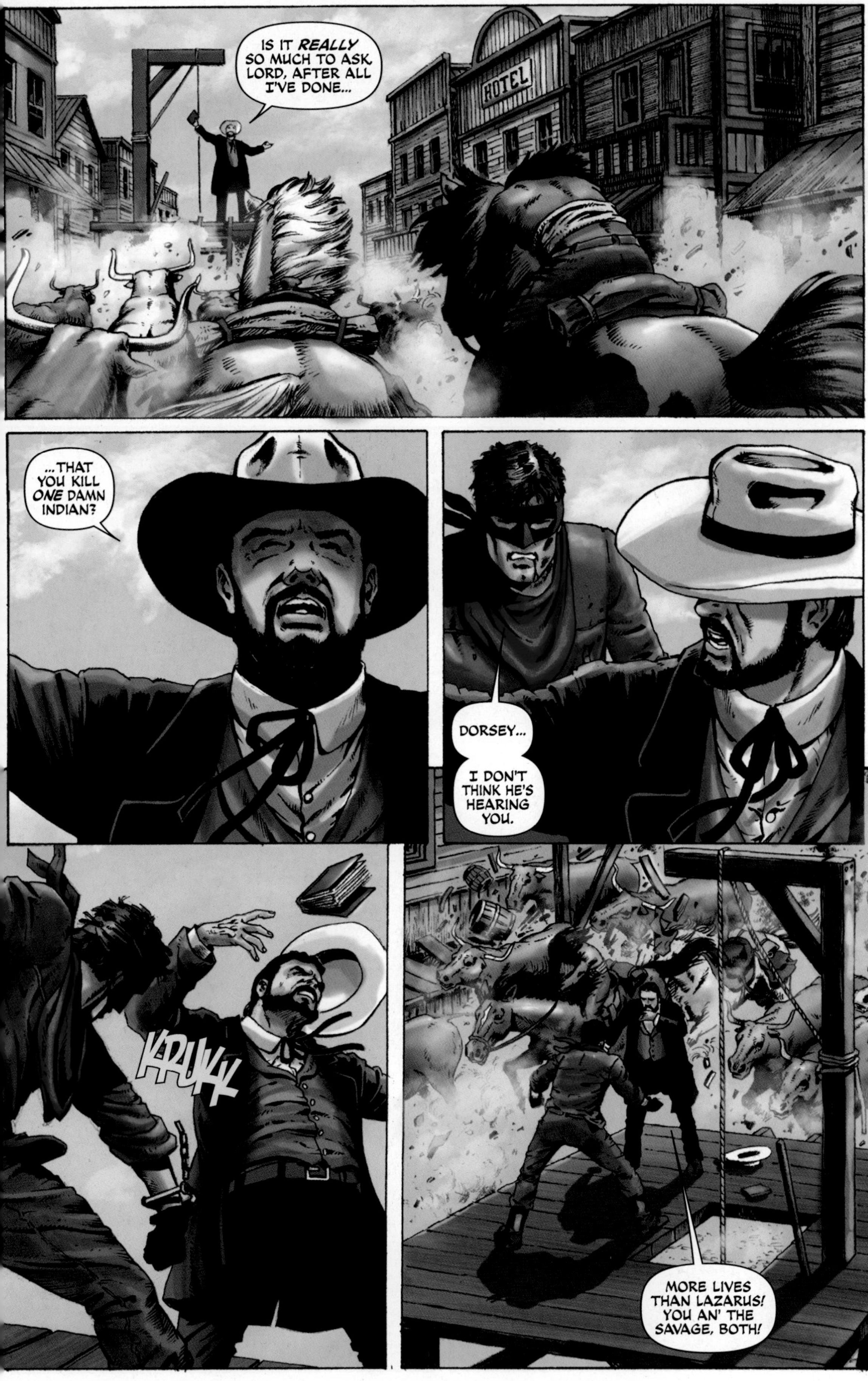
IS IT REALLY SO MUCH TO ASK, LORD, AFTER ALL I'VE DONE...
HOTEL
...THAT YOU KILL ONE DAMN INDIAN?
DORSEY...
I DON'T THINK HE'S HEARING YOU.
KRUKK
MORE LIVES THAN LAZARUS! YOU AN' THE SAVAGE, BOTH!

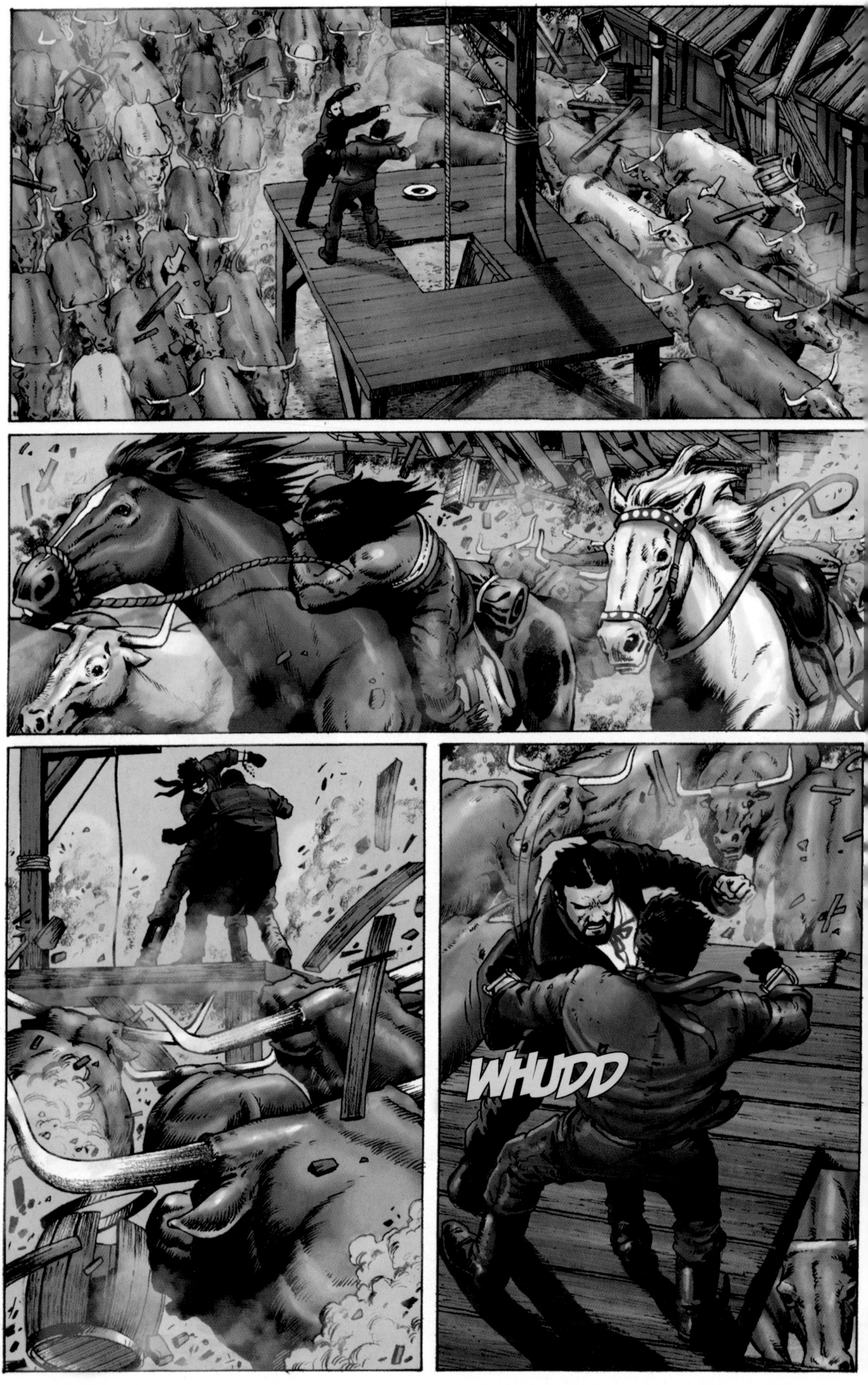
WHUDD

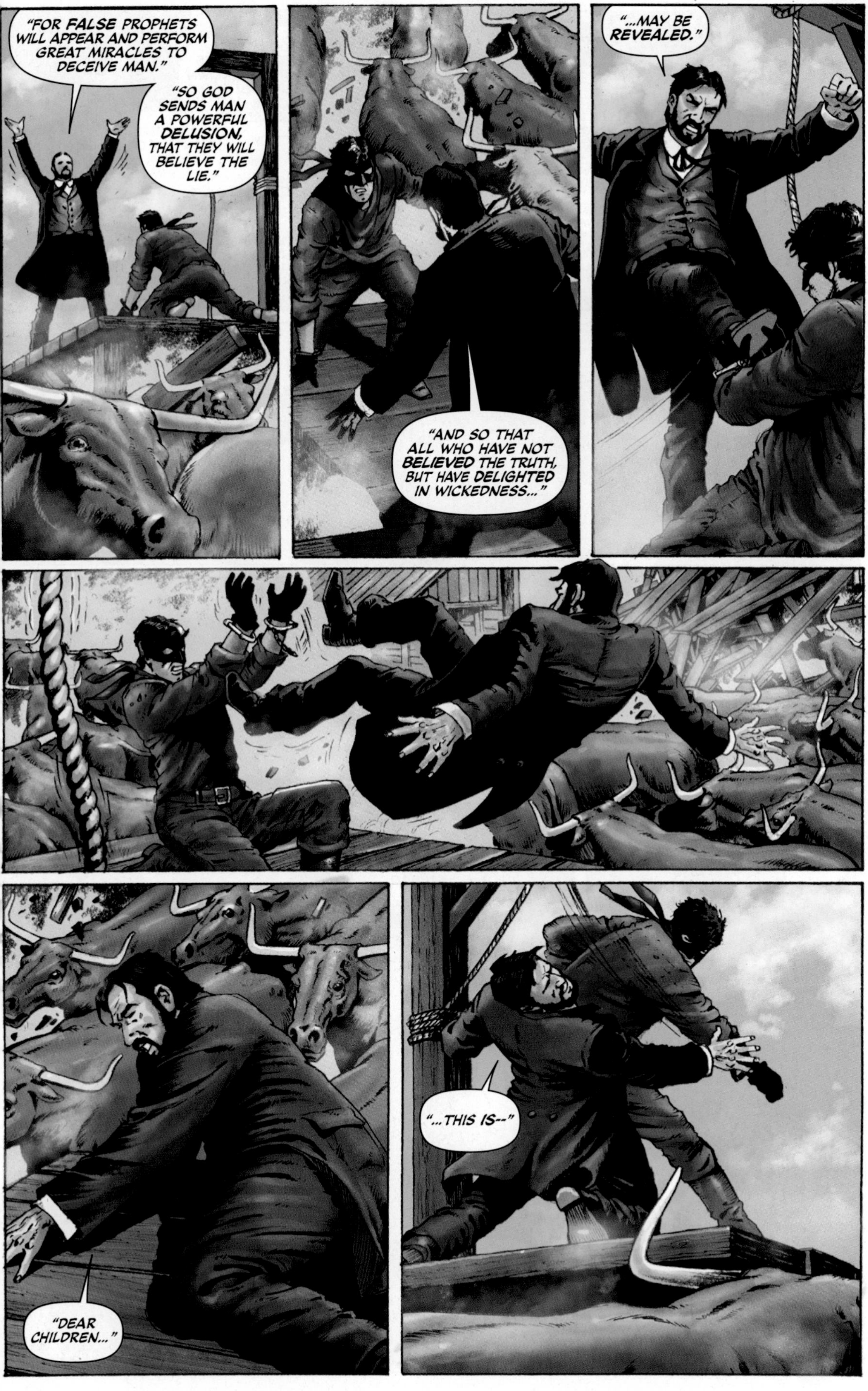
"FOR FALSE PROPHETS WILL APPEAR AND PERFORM GREAT MIRACLES TO DECEIVE MAN."
"SO GOD SENDS MAN A POWERFUL DELUSION, THAT THEY WILL BELIEVE THE LIE."
"AND SO THAT ALL WHO HAVE NOT BELIEVED THE TRUTH, BUT HAVE DELIGHTED IN WICKEDNESS..."
"...MAY BE REVEALED."
"DEAR CHILDREN..."
"...THIS IS--"

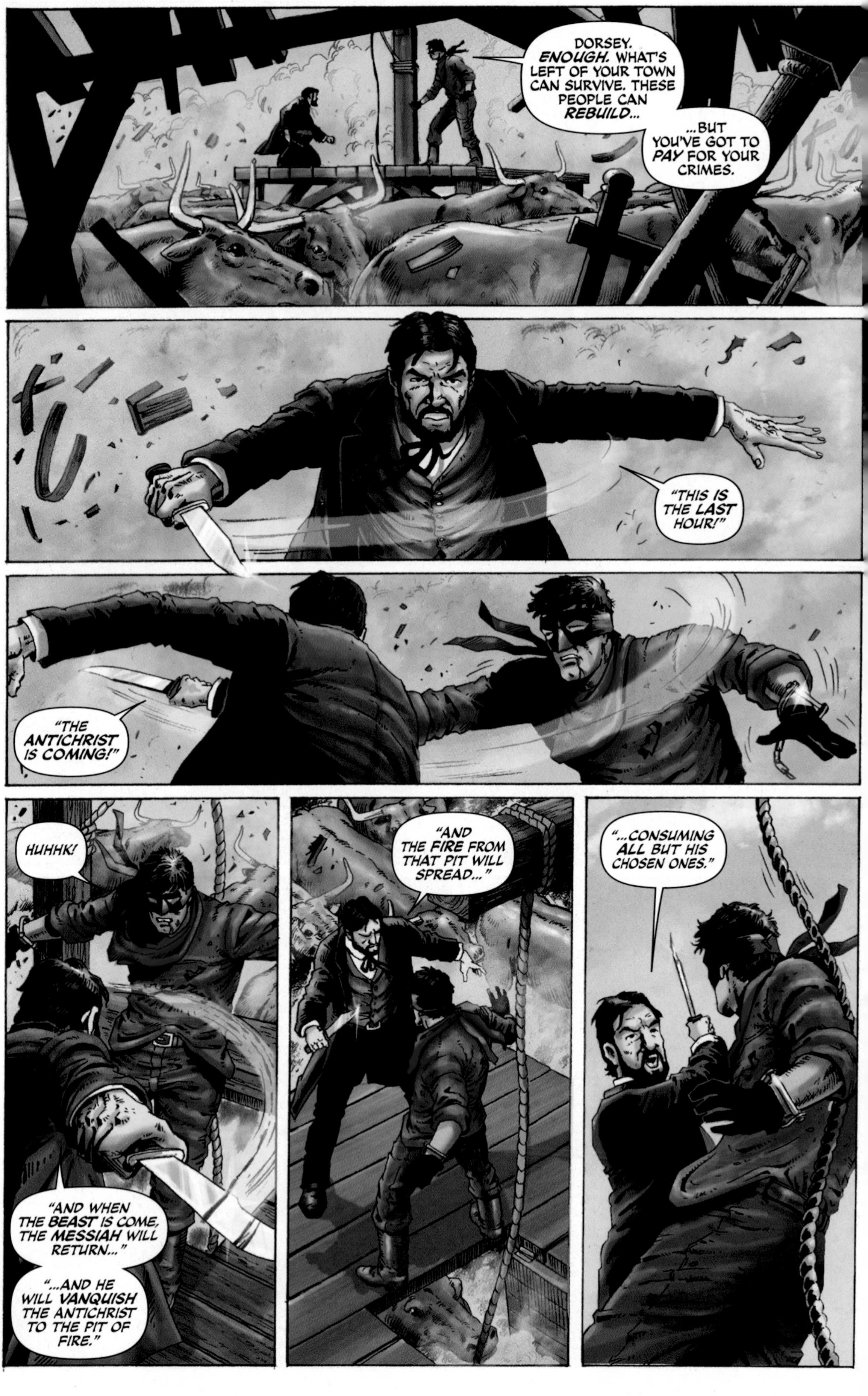
DORSEY. ENOUGH. WHAT'S LEFT OF YOUR TOWN CAN SURVIVE. THESE PEOPLE CAN REBUILD...
...BUT YOU'VE GOT TO PAY FOR YOUR CRIMES.
"THIS IS THE LAST HOUR!"
"THE ANTICHRIST IS COMING!"
HUHHK!
"AND WHEN THE BEAST IS COME, THE MESSIAH WILL RETURN..."
"...AND HE WILL VANQUISH THE ANTICHRIST TO THE PIT OF FIRE."
"AND THE FIRE FROM THAT PIT WILL SPREAD..."
"...CONSUMING ALL BUT HIS CHOSEN ONES."

"AND ALL OTHERS..."

"...SHALL BE CAST INTO THE PIT WITH THEIR *TRUE KING*...THE KING OF EVIL."

"AND HEAVEN ALONE SHALL WAIT..."

DRUMMBLLBRUMMDRUMMBLLBRUMM

DRUMMBLLBRUMM

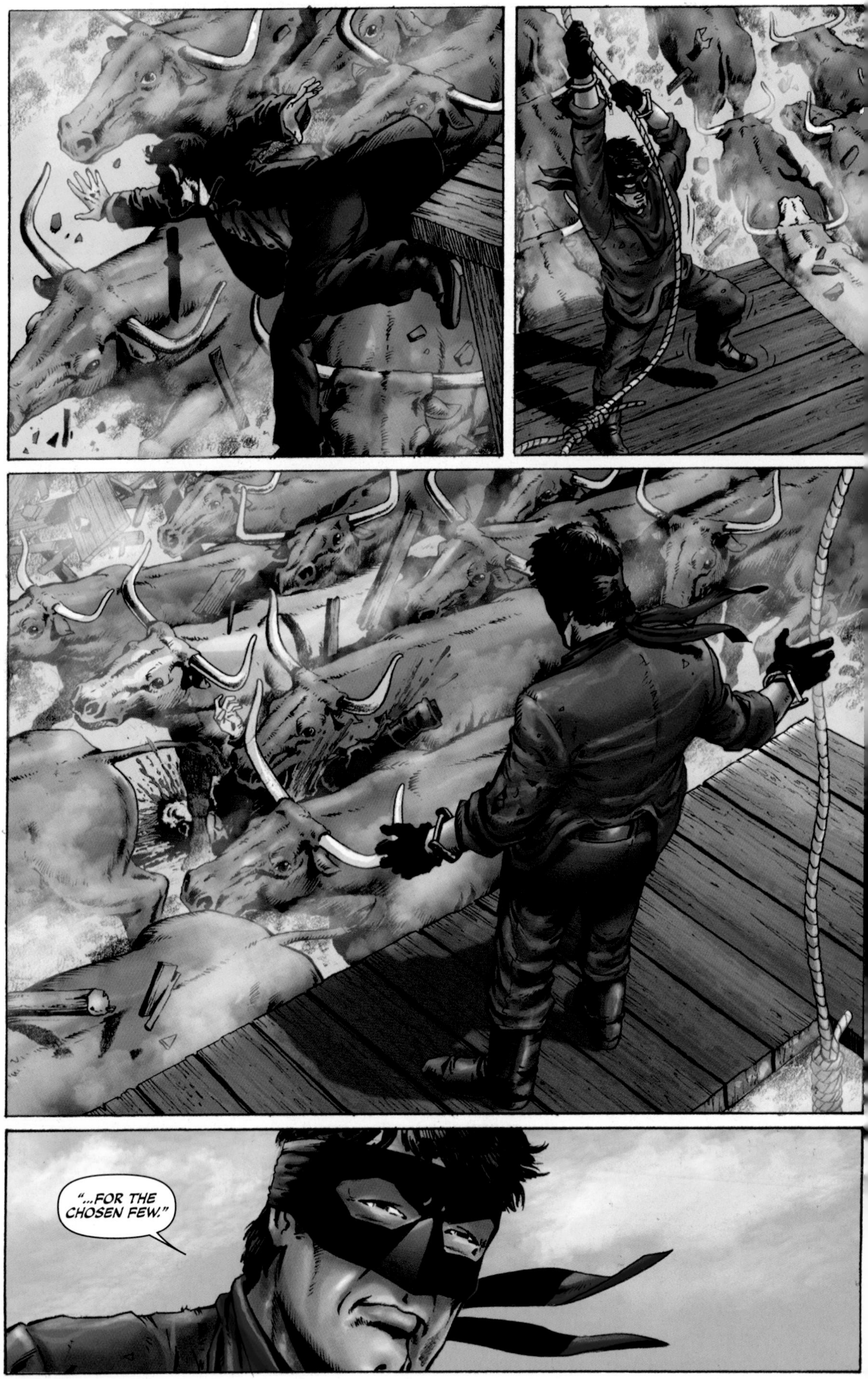
"...FOR THE CHOSEN FEW."

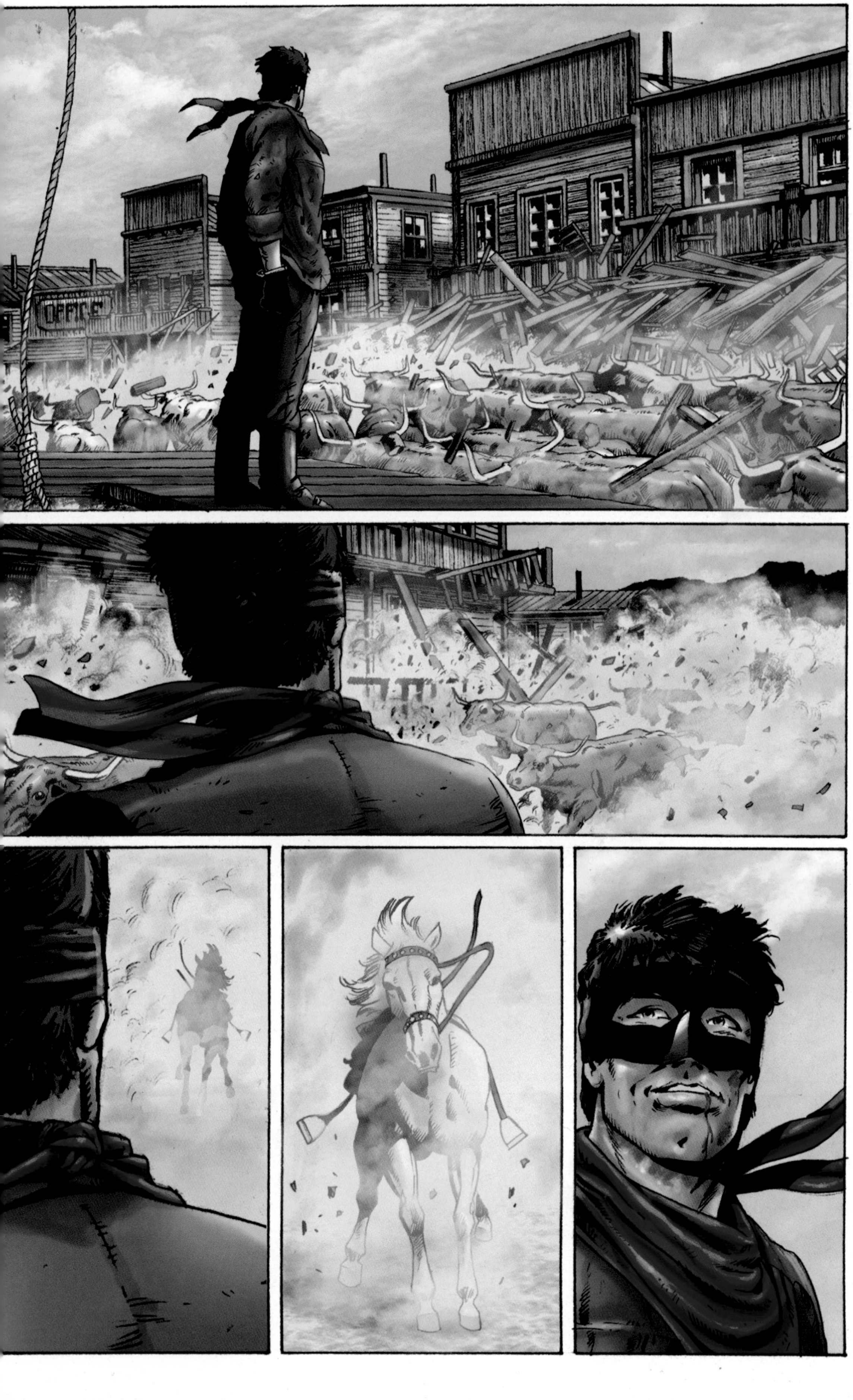
OFFICE

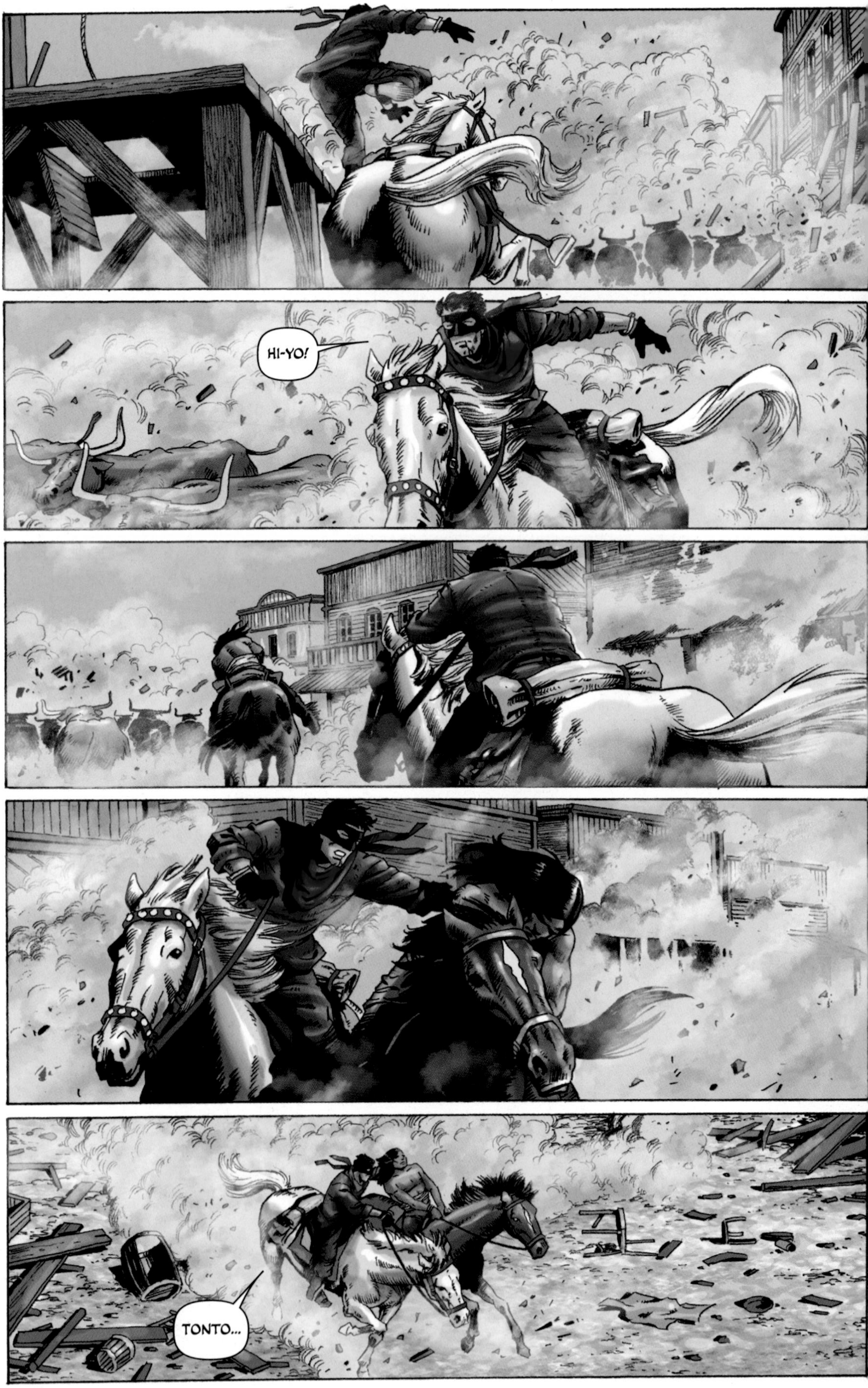
HI-YO!
TONTO...

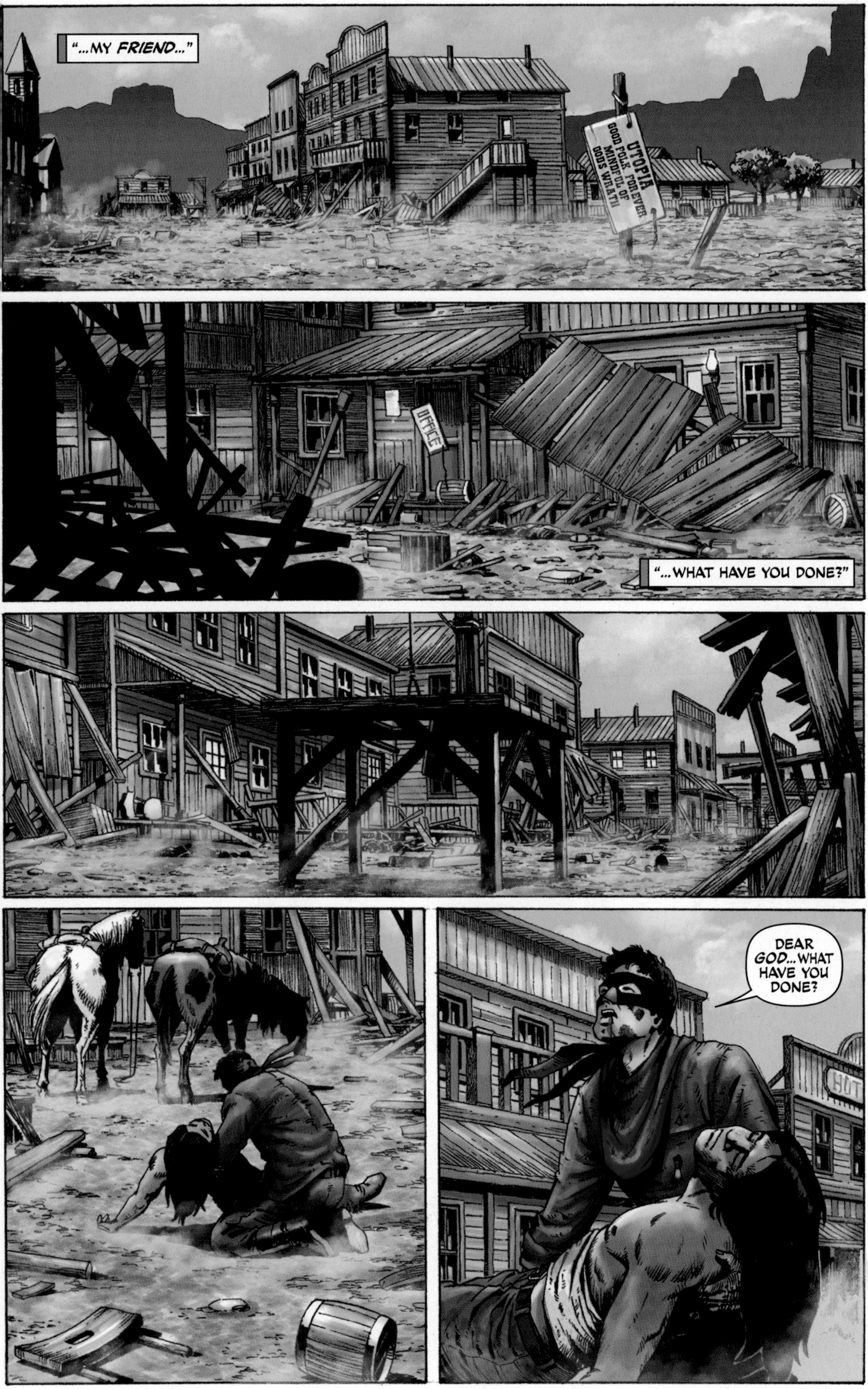
"...MY FRIEND..."
UTOPIA GOOD FOLK FOREVER MINDFUL OF GODS WRATH
OFFICE
"...WHAT HAVE YOU DONE?"
DEAR GOD...WHAT HAVE YOU DONE?

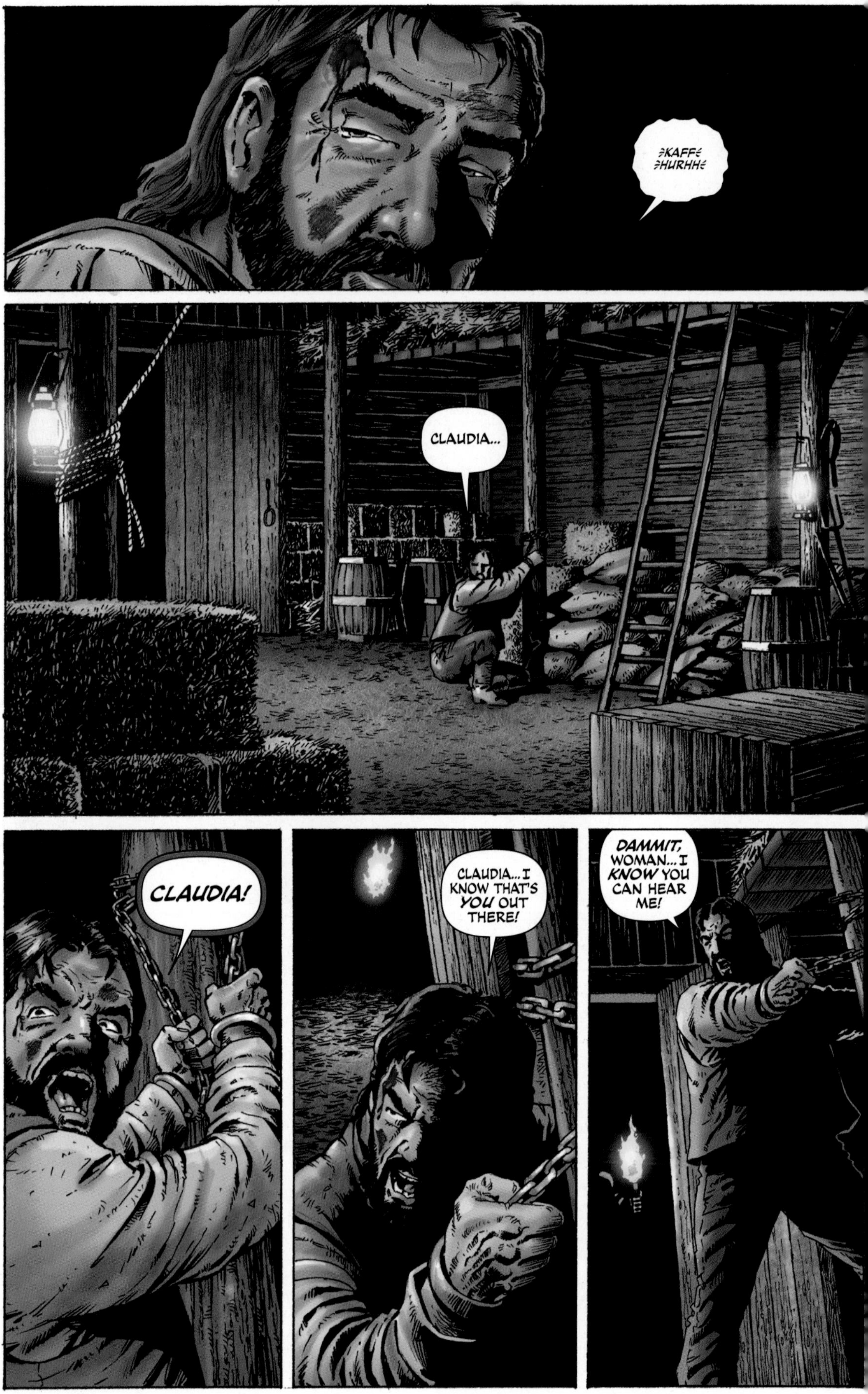

≷KAFF≶
≷HURHH≶
CLAUDIA...
CLAUDIA!
CLAUDIA... I KNOW THAT'S *YOU* OUT THERE!
DAMMIT, WOMAN... I ***KNOW*** YOU CAN HEAR ME!

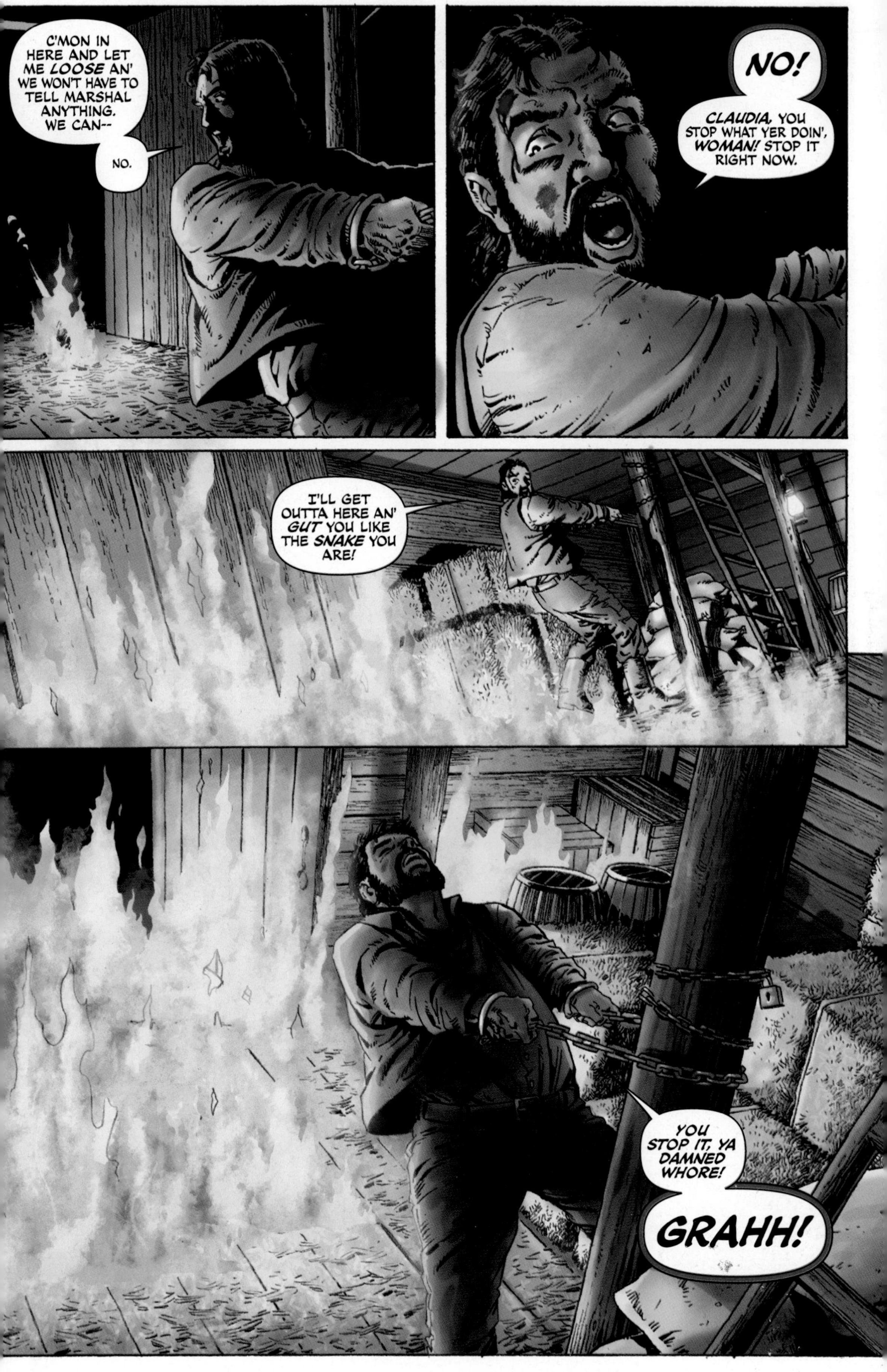
C'MON IN HERE AND LET ME LOOSE AN' WE WON'T HAVE TO TELL MARSHAL ANYTHING. WE CAN--
NO.
NO!
CLAUDIA, YOU STOP WHAT YER DOIN', WOMAN! STOP IT RIGHT NOW.
I'LL GET OUTTA HERE AN' GUT YOU LIKE THE SNAKE YOU ARE!
YOU STOP IT, YA DAMNED WHORE!
GRAHH!

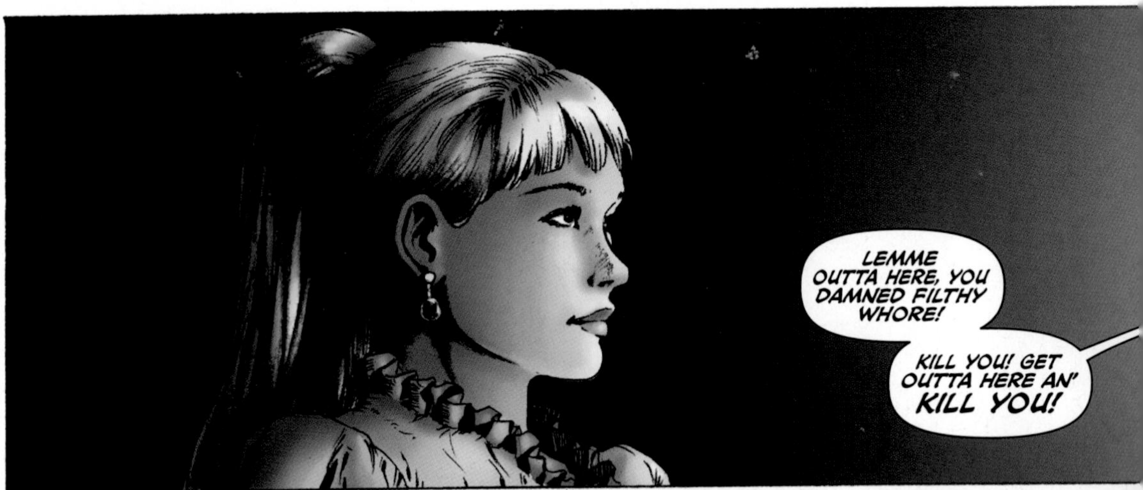
LEMME OUTTA HERE, YOU DAMNED FILTHY WHORE!
KILL YOU! GET OUTTA HERE AN' KILL YOU!

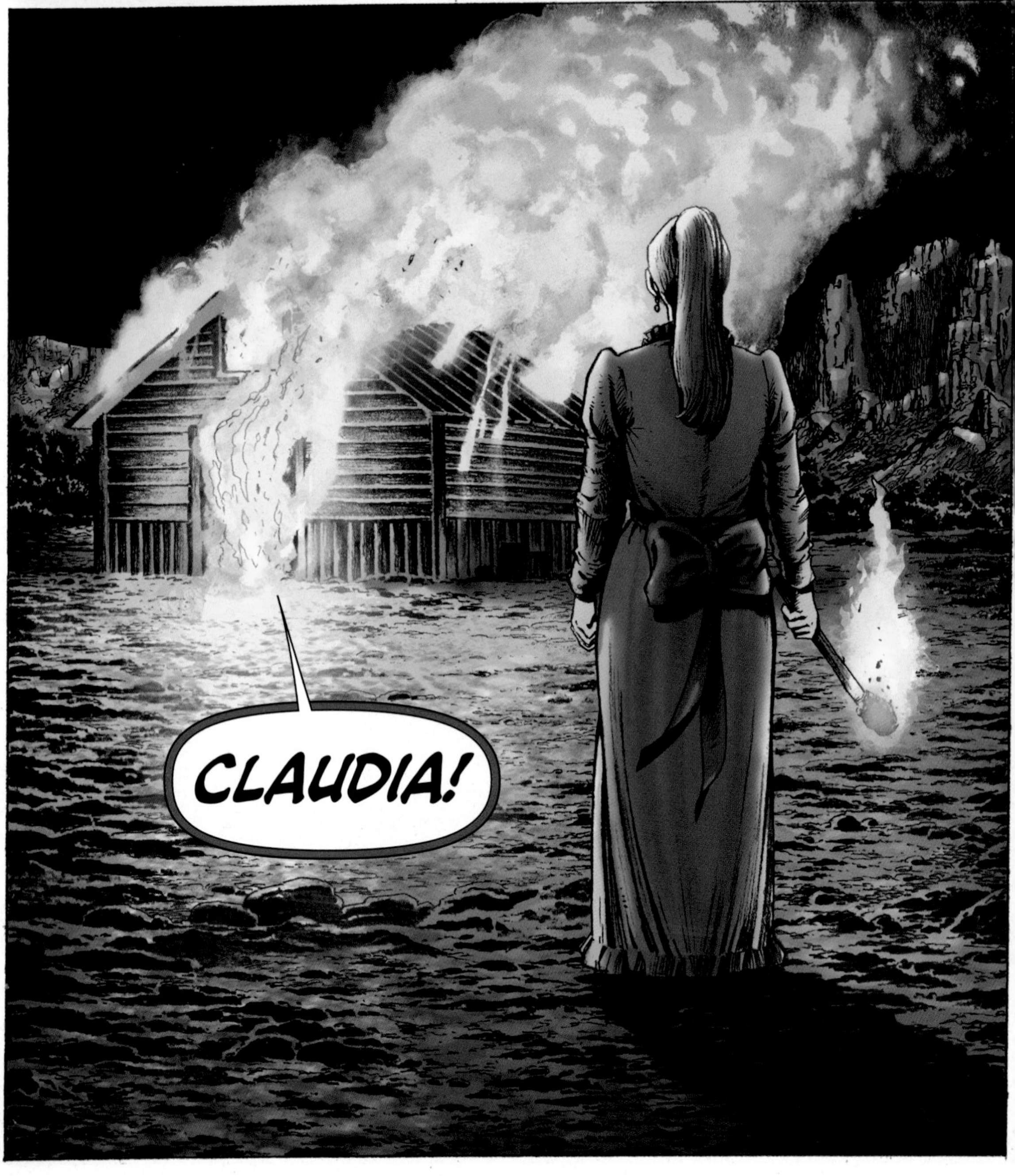
CLAUDIA!

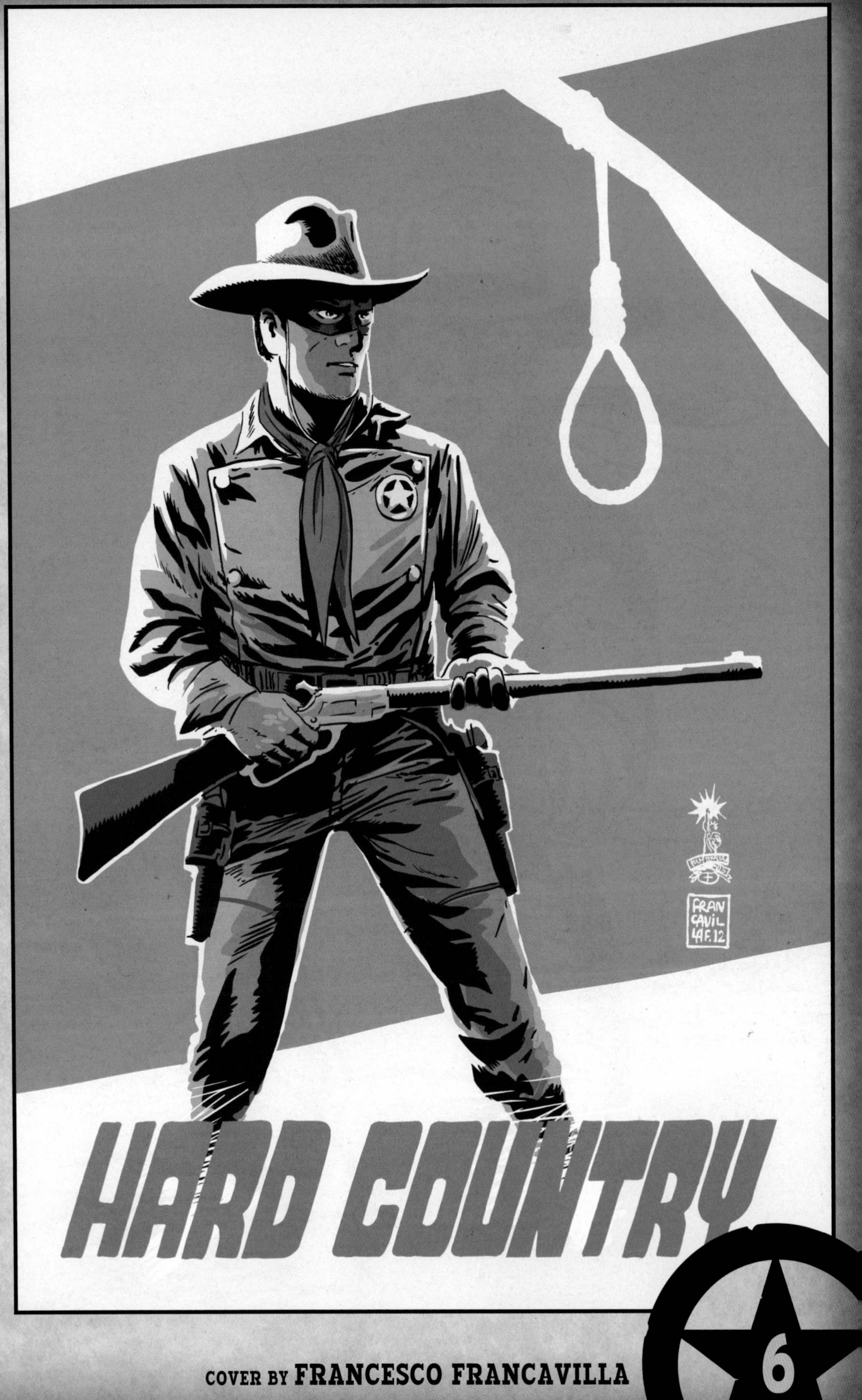
HARD COUNTRY
FRANCAVILLA F.12
6

COVER BY **FRANCESCO FRANCAVILLA**

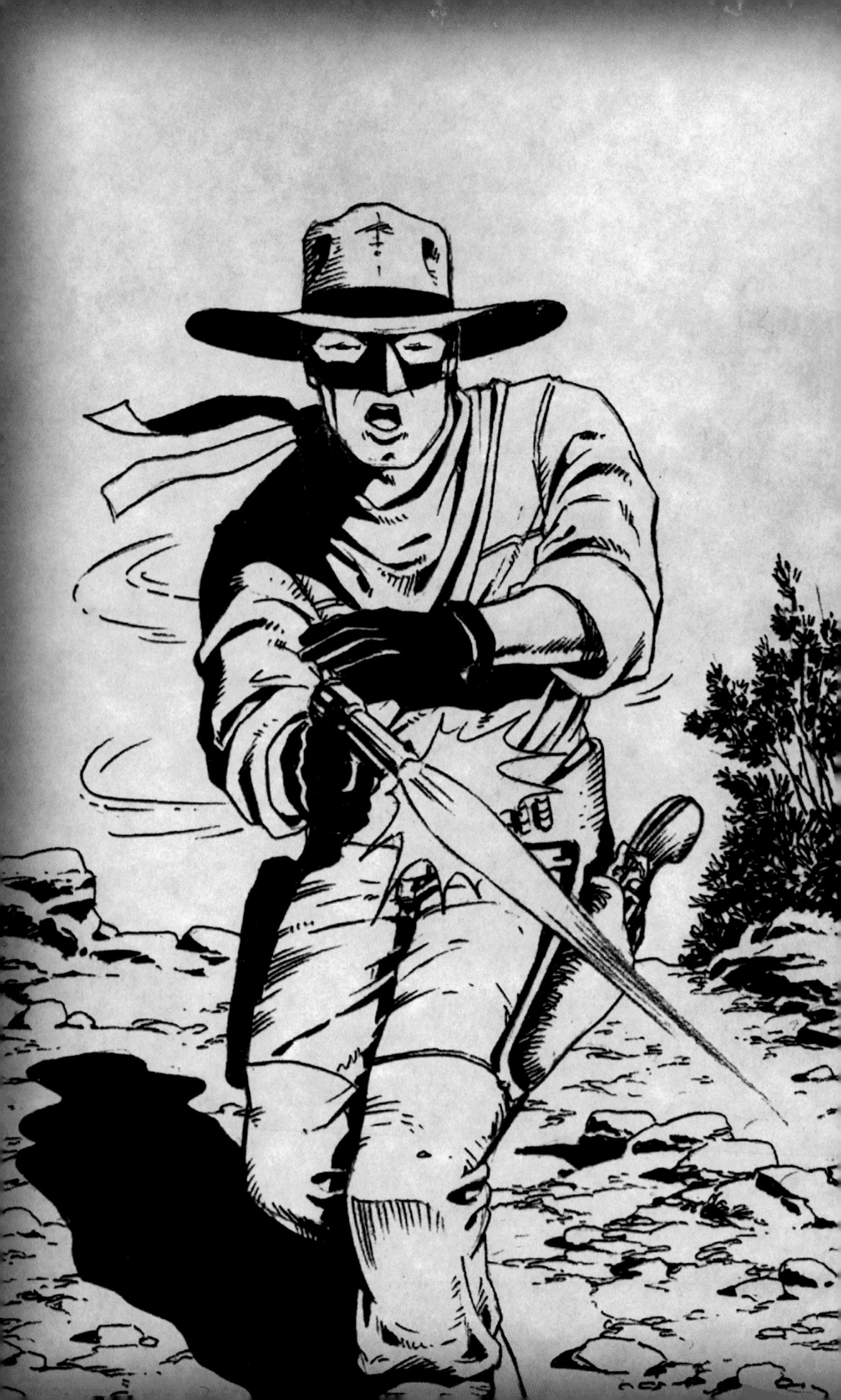

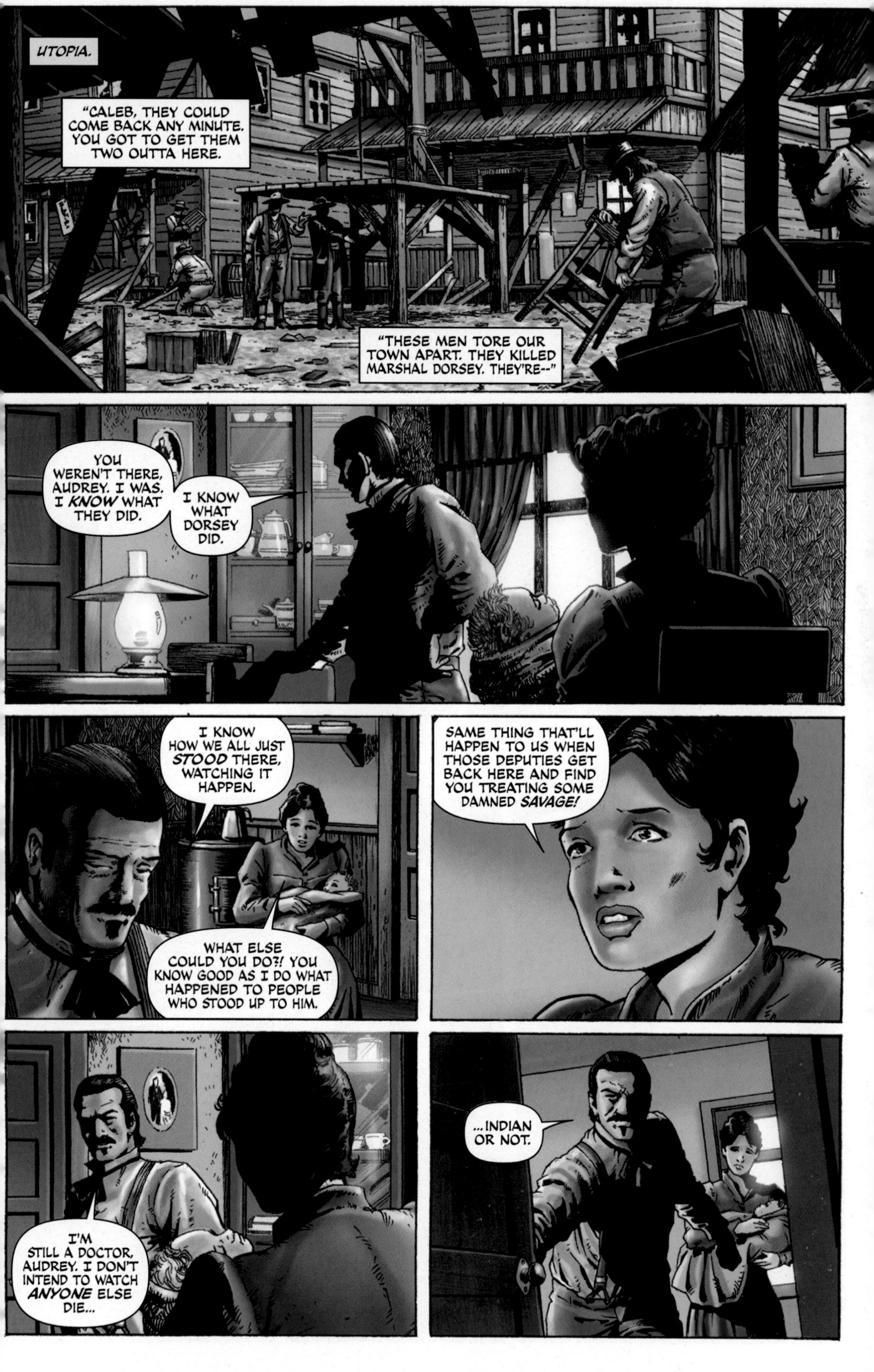
UTOPIA.
"CALEB, THEY COULD COME BACK ANY MINUTE. YOU GOT TO GET THEM TWO OUTTA HERE.
"THESE MEN TORE OUR TOWN APART. THEY KILLED MARSHAL DORSEY. THEY'RE--"
YOU WEREN'T THERE, AUDREY. I WAS. I KNOW WHAT THEY DID.
I KNOW WHAT DORSEY DID.
I KNOW HOW WE ALL JUST STOOD THERE, WATCHING IT HAPPEN.
WHAT ELSE COULD YOU DO?! YOU KNOW GOOD AS I DO WHAT HAPPENED TO PEOPLE WHO STOOD UP TO HIM.
SAME THING THAT'LL HAPPEN TO US WHEN THOSE DEPUTIES GET BACK HERE AND FIND YOU TREATING SOME DAMNED SAVAGE!
I'M STILL A DOCTOR, AUDREY. I DON'T INTEND TO WATCH ANYONE ELSE DIE...
...INDIAN OR NOT.

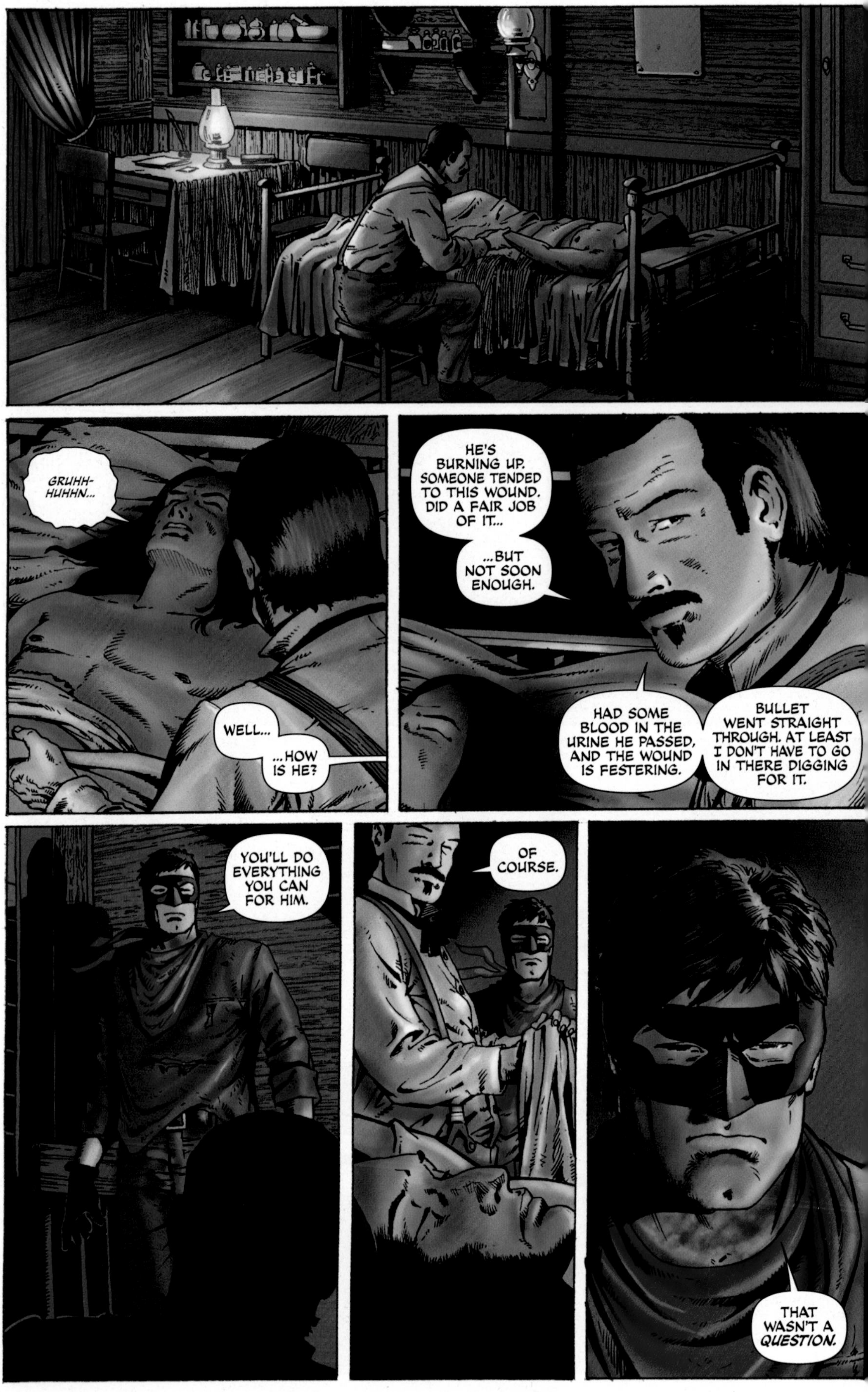
GRUHH-HUHHN...
WELL...
...HOW IS HE?
HE'S BURNING UP. SOMEONE TENDED TO THIS WOUND. DID A FAIR JOB OF IT...
...BUT NOT SOON ENOUGH.
BULLET WENT STRAIGHT THROUGH. AT LEAST I DON'T HAVE TO GO IN THERE DIGGING FOR IT.
HAD SOME BLOOD IN THE URINE HE PASSED, AND THE WOUND IS FESTERING.
YOU'LL DO EVERYTHING YOU CAN FOR HIM.
OF COURSE.
THAT WASN'T A *QUESTION.*

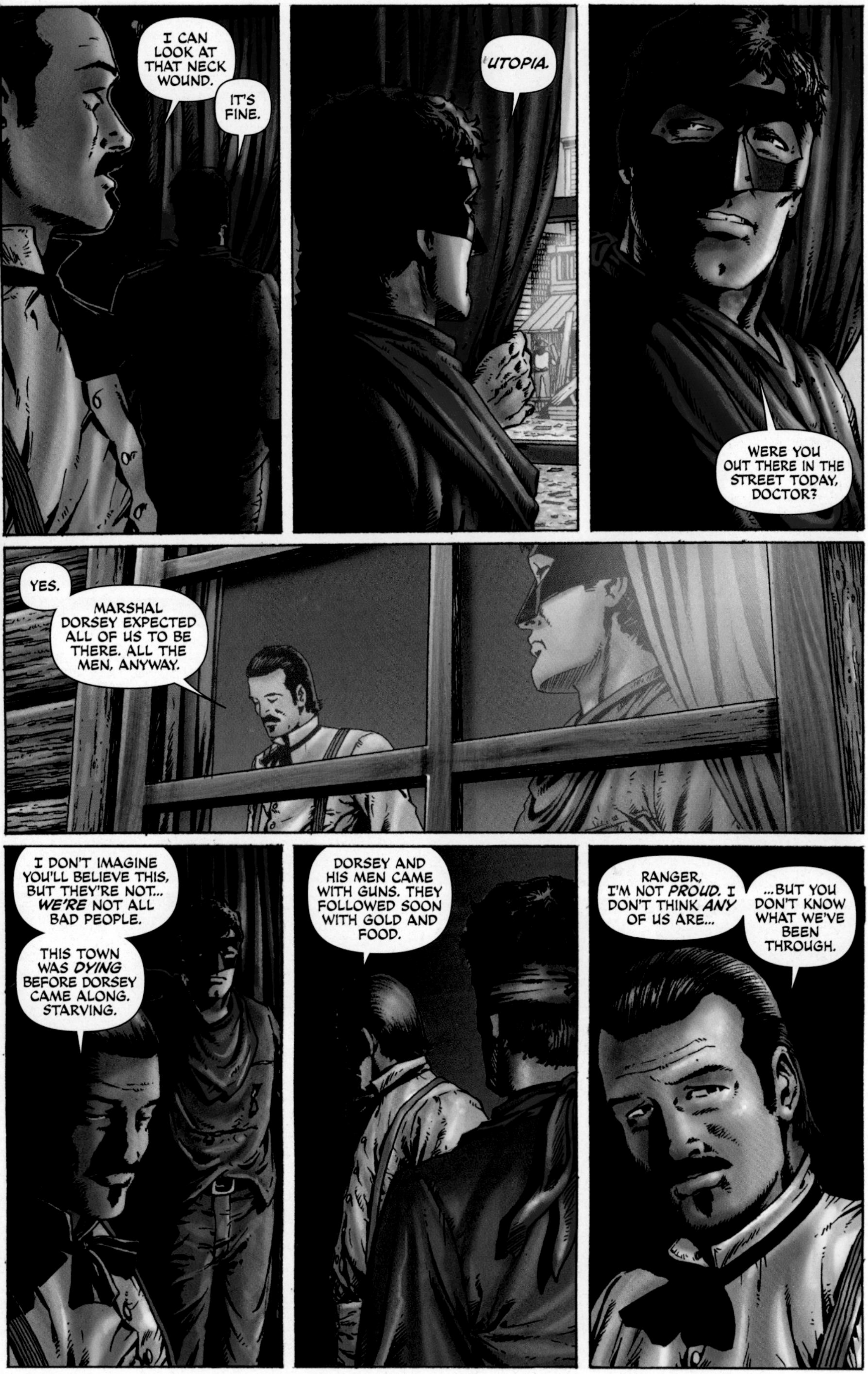
I CAN LOOK AT THAT NECK WOUND.
IT'S FINE.
UTOPIA.
WERE YOU OUT THERE IN THE STREET TODAY, DOCTOR?
YES.
MARSHAL DORSEY EXPECTED ALL OF US TO BE THERE. ALL THE MEN, ANYWAY.
I DON'T IMAGINE YOU'LL BELIEVE THIS, BUT THEY'RE NOT... *WE'RE* NOT ALL BAD PEOPLE.
THIS TOWN WAS *DYING* BEFORE DORSEY CAME ALONG. STARVING.
DORSEY AND HIS MEN CAME WITH GUNS. THEY FOLLOWED SOON WITH GOLD AND FOOD.
RANGER, I'M NOT *PROUD.* I DON'T THINK *ANY* OF US ARE...
...BUT YOU DON'T KNOW WHAT WE'VE BEEN THROUGH.

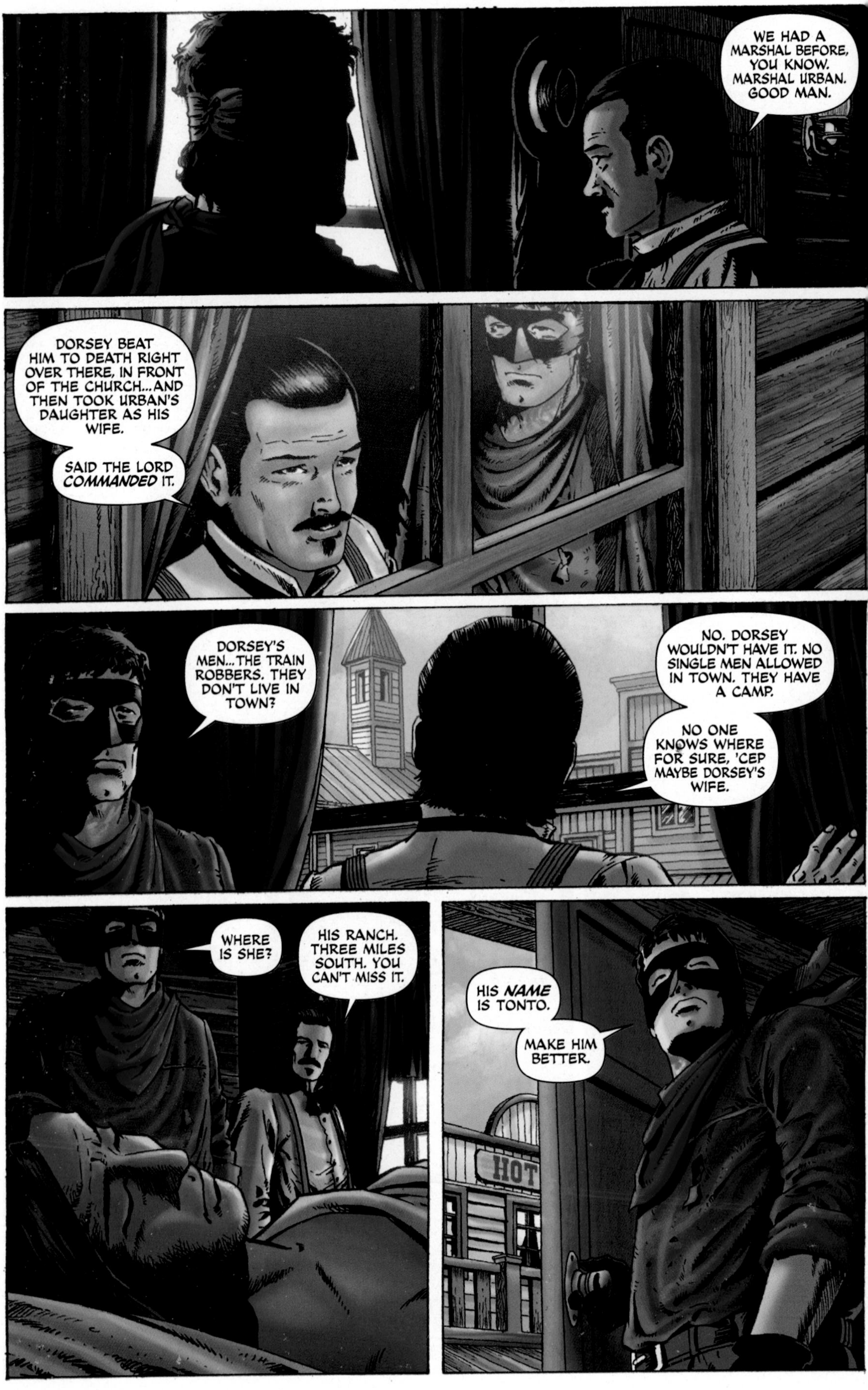
WE HAD A MARSHAL BEFORE, YOU KNOW. MARSHAL URBAN. GOOD MAN.
DORSEY BEAT HIM TO DEATH RIGHT OVER THERE, IN FRONT OF THE CHURCH...AND THEN TOOK URBAN'S DAUGHTER AS HIS WIFE.
SAID THE LORD *COMMANDED* IT.
DORSEY'S MEN...THE TRAIN ROBBERS. THEY DON'T LIVE IN TOWN?
NO. DORSEY WOULDN'T HAVE IT. NO SINGLE MEN ALLOWED IN TOWN. THEY HAVE A CAMP.
NO ONE KNOWS WHERE FOR SURE, 'CEP MAYBE DORSEY'S WIFE.
WHERE IS SHE?
HIS RANCH. THREE MILES SOUTH. YOU CAN'T MISS IT.
HIS *NAME* IS TONTO.
MAKE HIM BETTER.
HOT

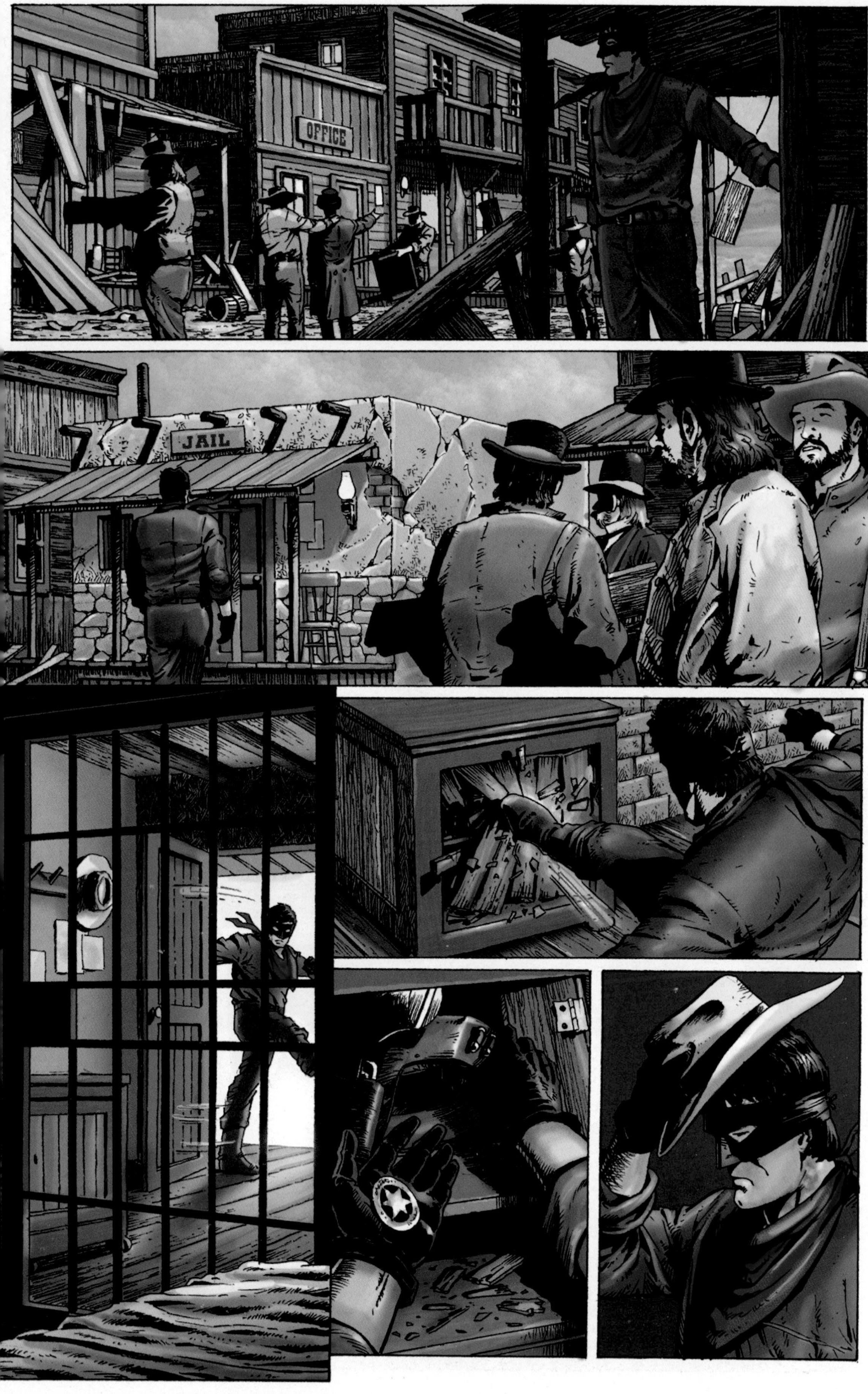
OFFICE
JAIL

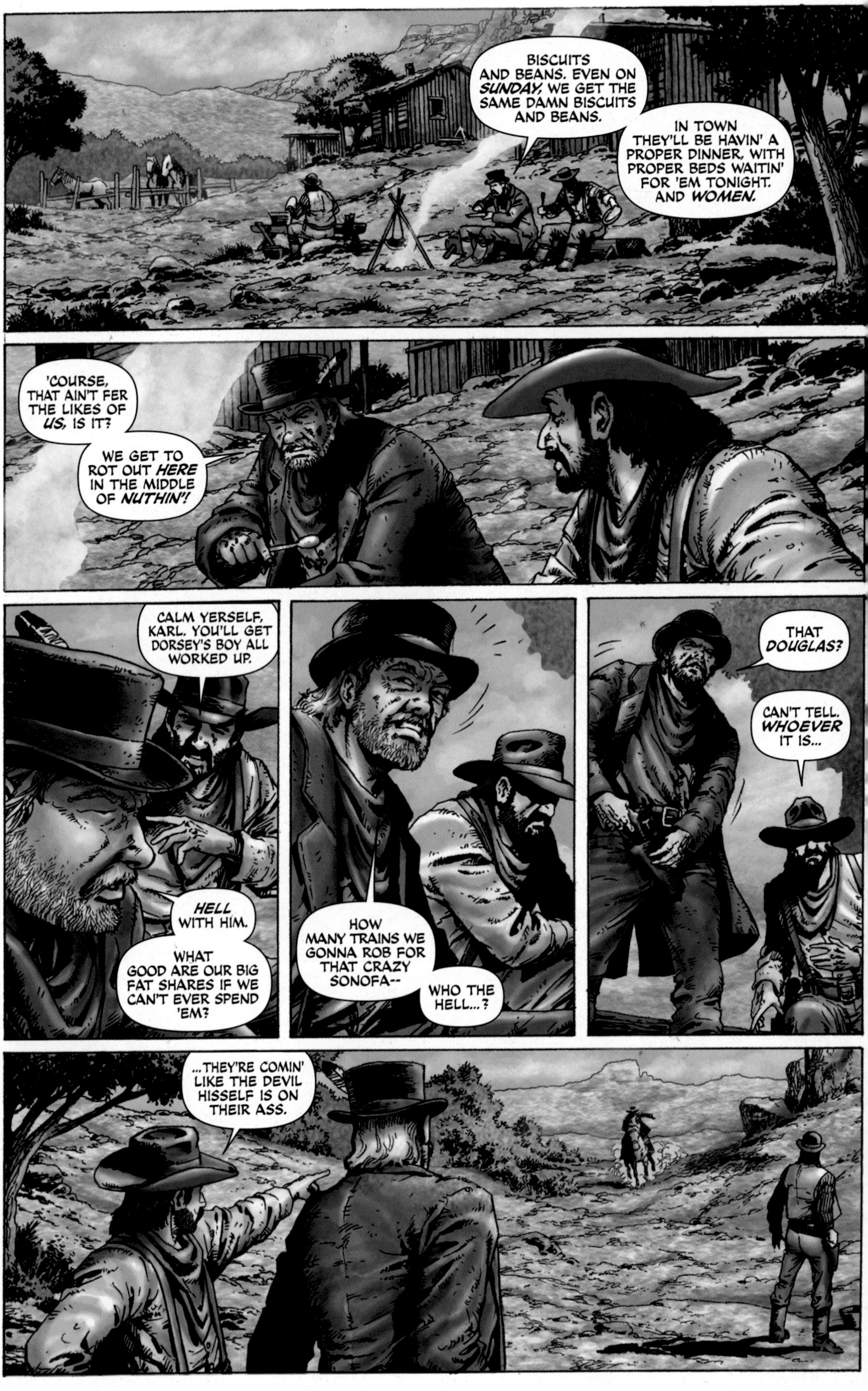
BISCUITS AND BEANS. EVEN ON SUNDAY, WE GET THE SAME DAMN BISCUITS AND BEANS.
IN TOWN THEY'LL BE HAVIN' A PROPER DINNER, WITH PROPER BEDS WAITIN' FOR 'EM TONIGHT. AND WOMEN.
'COURSE, THAT AIN'T FER THE LIKES OF US, IS IT?
WE GET TO ROT OUT HERE IN THE MIDDLE OF NUTHIN'!
CALM YERSELF, KARL. YOU'LL GET DORSEY'S BOY ALL WORKED UP.
HELL WITH HIM.
WHAT GOOD ARE OUR BIG FAT SHARES IF WE CAN'T EVER SPEND 'EM?
HOW MANY TRAINS WE GONNA ROB FOR THAT CRAZY SONOFA--
WHO THE HELL...?
THAT DOUGLAS?
CAN'T TELL. WHOEVER IT IS...
...THEY'RE COMIN' LIKE THE DEVIL HISSELF IS ON THEIR ASS.

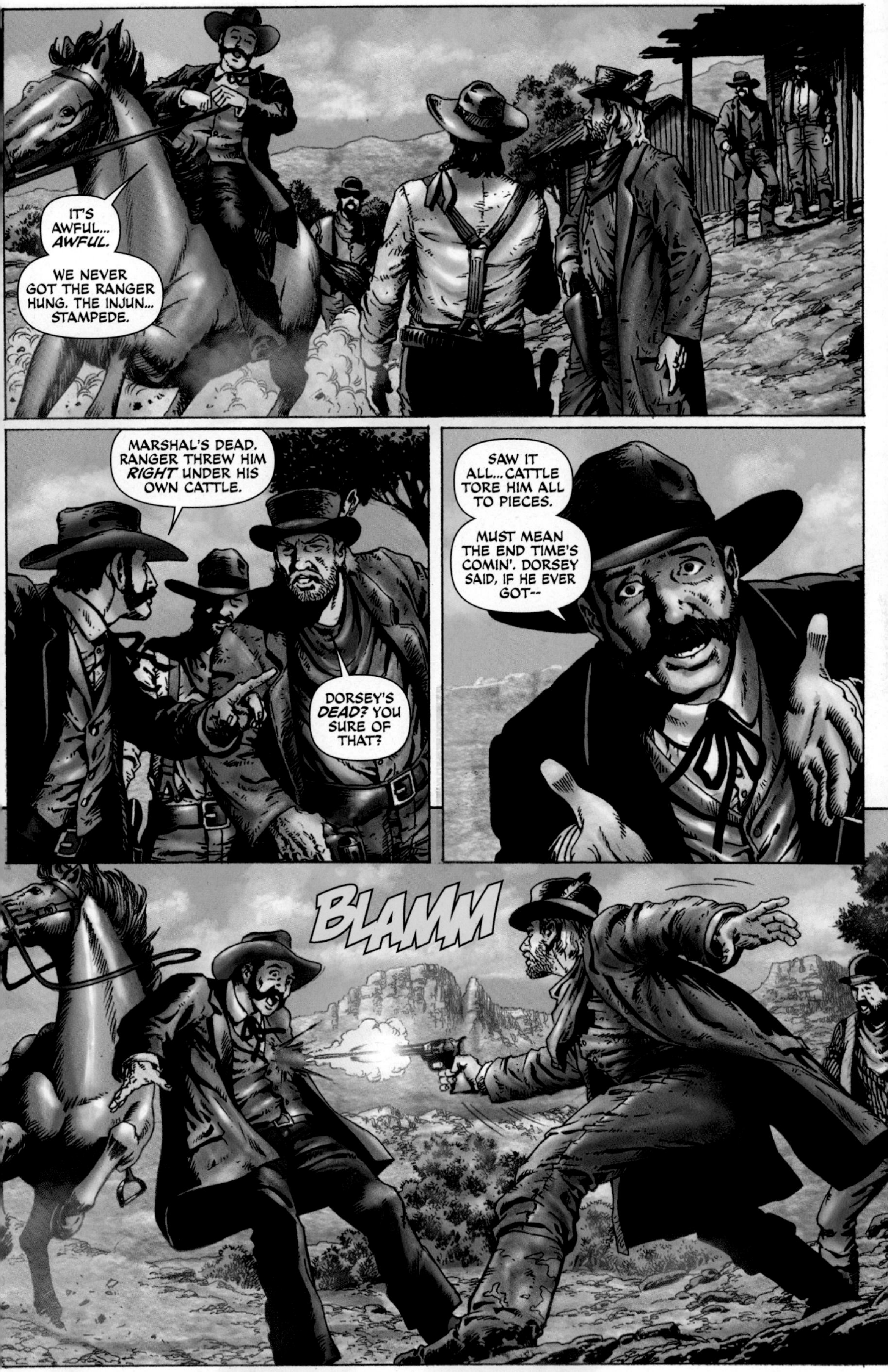
IT'S AWFUL... *AWFUL.*
WE NEVER GOT THE RANGER HUNG. THE INJUN... STAMPEDE.
MARSHAL'S DEAD. RANGER THREW HIM *RIGHT* UNDER HIS OWN CATTLE.
DORSEY'S *DEAD?* YOU SURE OF THAT?
SAW IT ALL...CATTLE TORE HIM ALL TO PIECES.
MUST MEAN THE END TIME'S COMIN'. DORSEY SAID, IF HE EVER GOT--
BLAMM

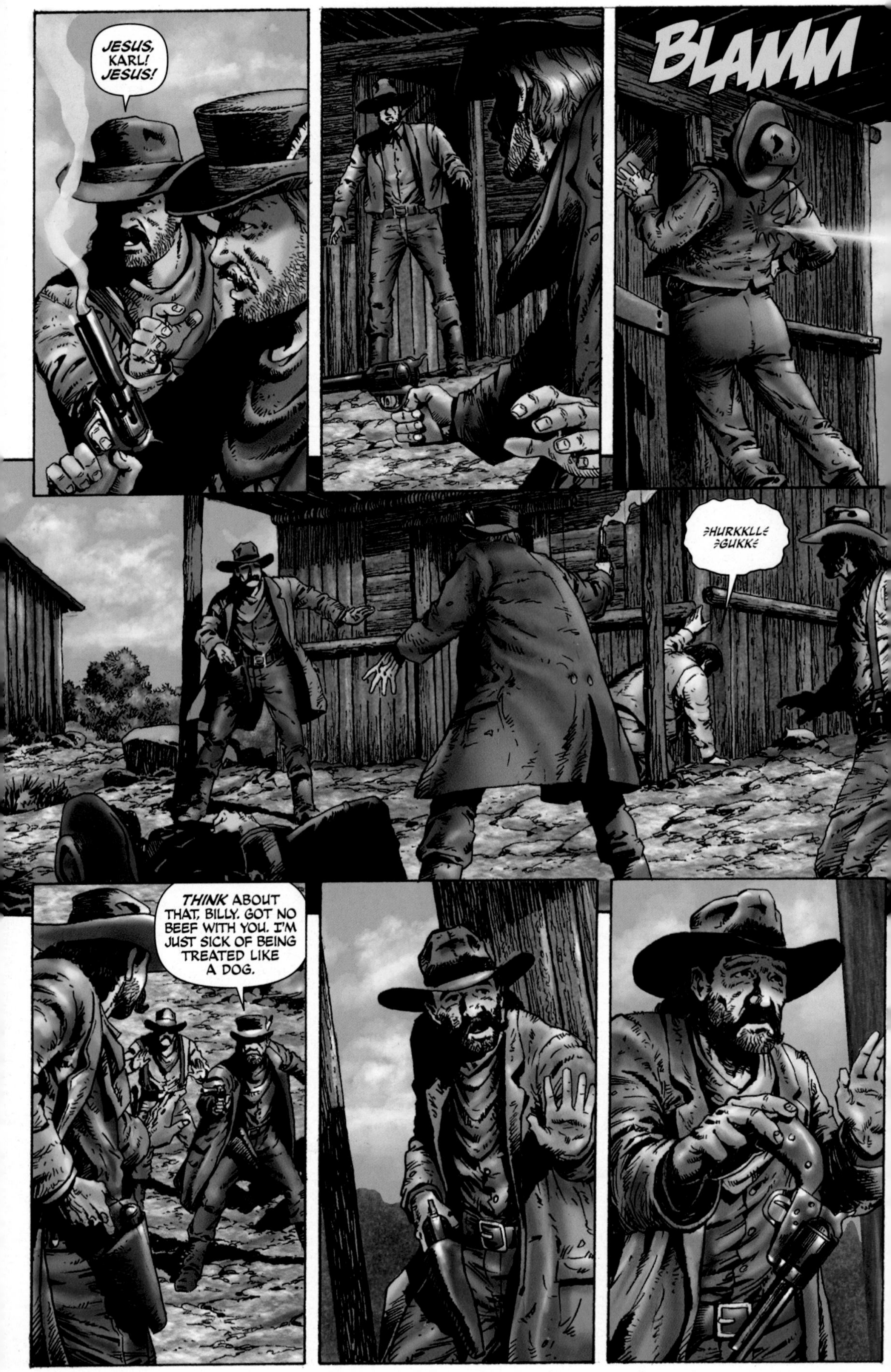
JESUS, KARL! JESUS!
BLAMM
≠HURKKLL≠ ≠GUKK≠
THINK ABOUT THAT, BILLY. GOT NO BEEF WITH YOU. I'M JUST SICK OF BEING TREATED LIKE A DOG.

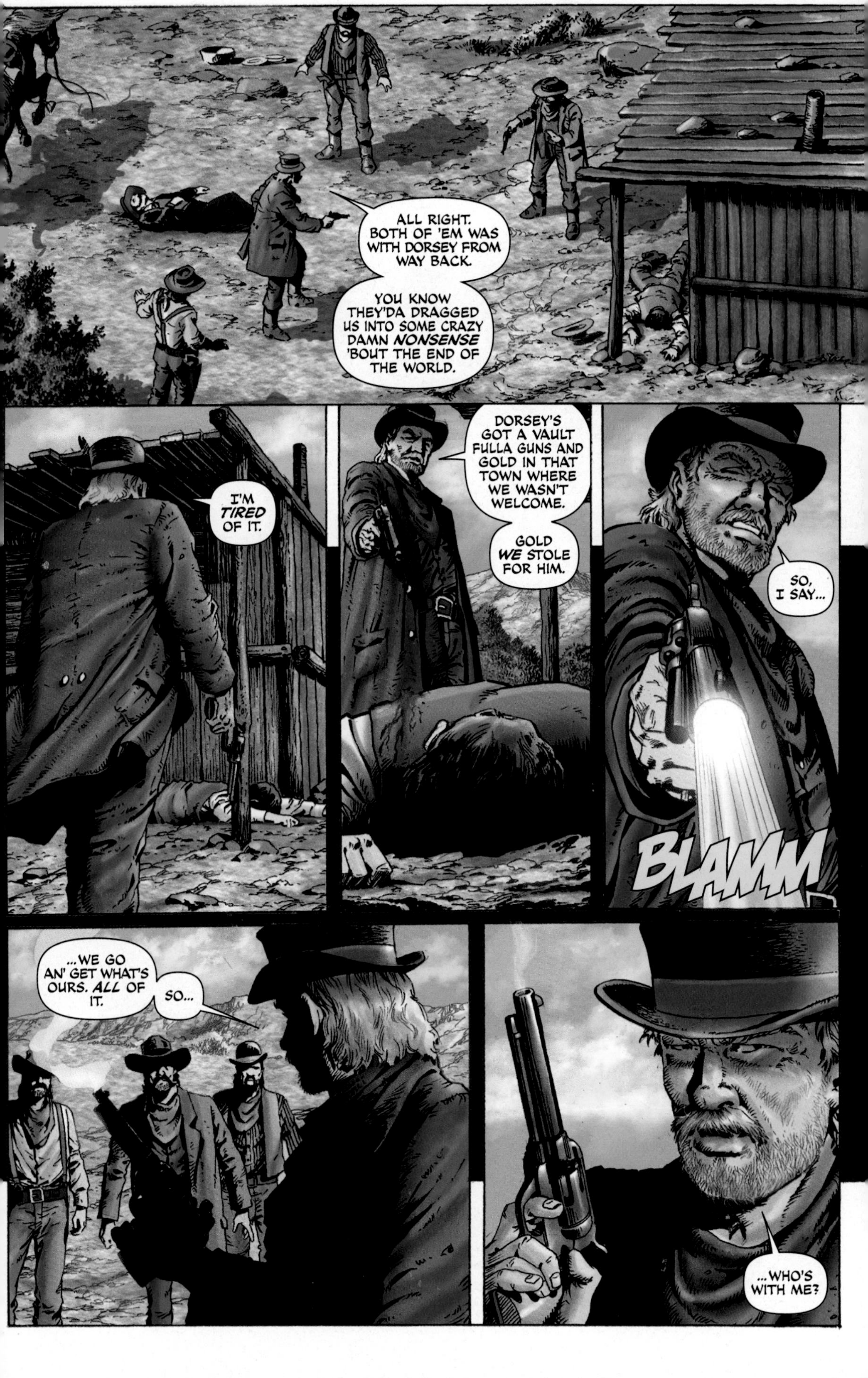
ALL RIGHT. BOTH OF 'EM WAS WITH DORSEY FROM WAY BACK.
YOU KNOW THEY'DA DRAGGED US INTO SOME CRAZY DAMN *NONSENSE* 'BOUT THE END OF THE WORLD.
I'M *TIRED* OF IT.
DORSEY'S GOT A VAULT FULLA GUNS AND GOLD IN THAT TOWN WHERE WE WASN'T WELCOME.
GOLD *WE* STOLE FOR HIM.
SO, I SAY...
BLAMM
...WE GO AN' GET WHAT'S OURS. *ALL* OF IT.
SO...
...WHO'S WITH ME?

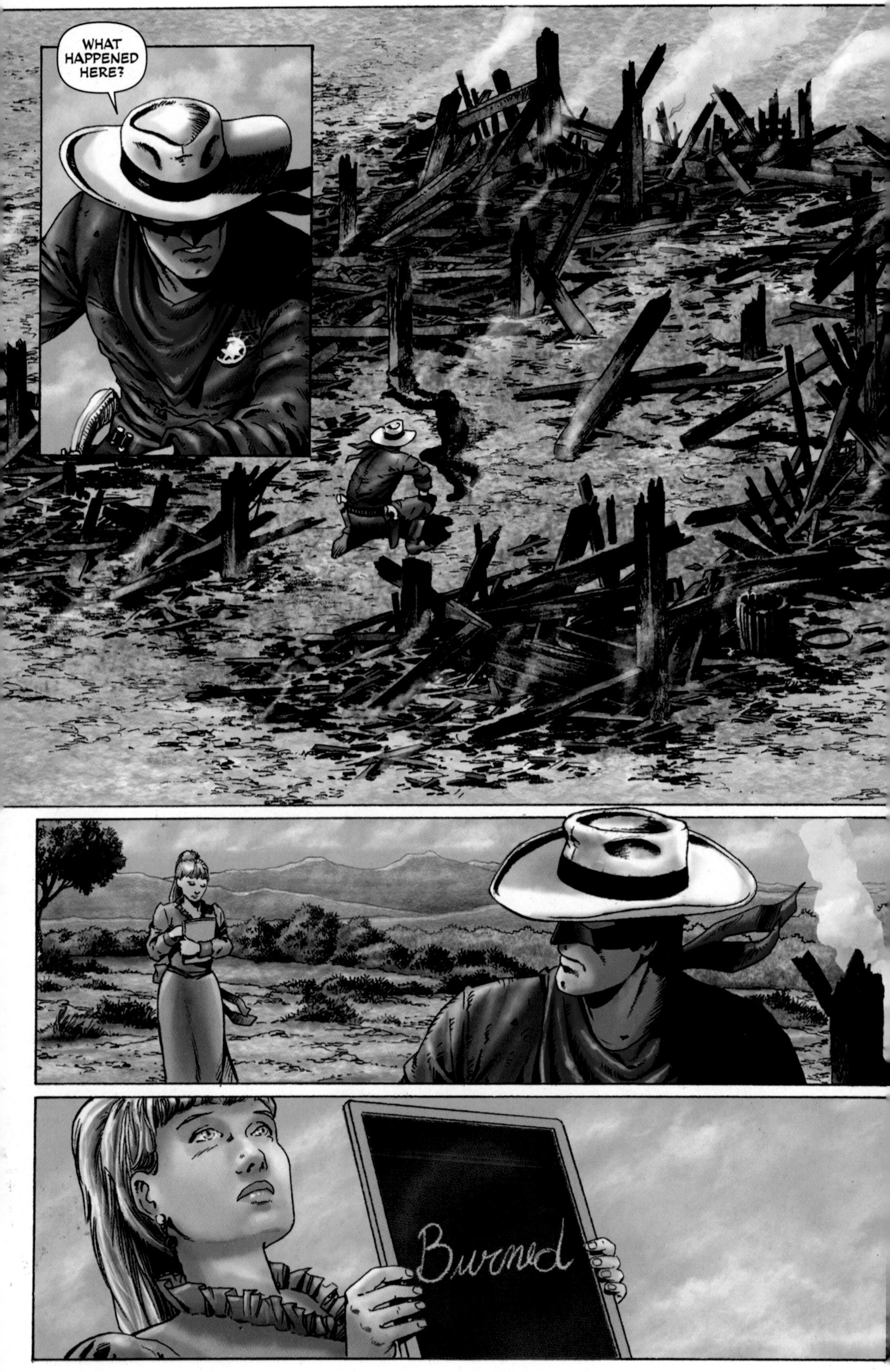
WHAT HAPPENED HERE?
Burned

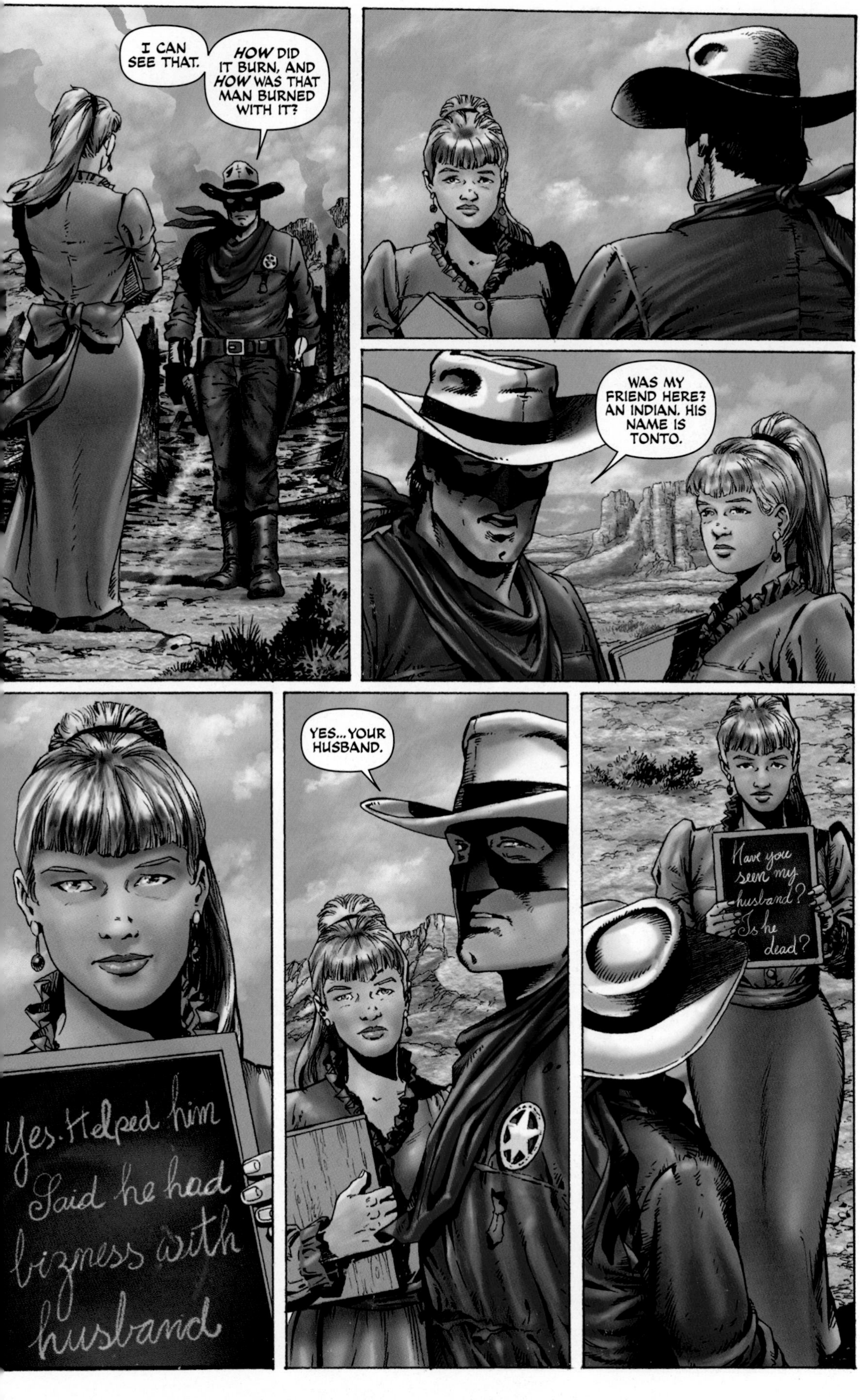
I CAN SEE THAT.
HOW DID IT BURN, AND *HOW* WAS THAT MAN BURNED WITH IT?
WAS MY FRIEND HERE? AN INDIAN. HIS NAME IS TONTO.
Yes. Helped him Said he had bizness with husband
YES...YOUR HUSBAND.
Have you seen my husband? Is he dead?

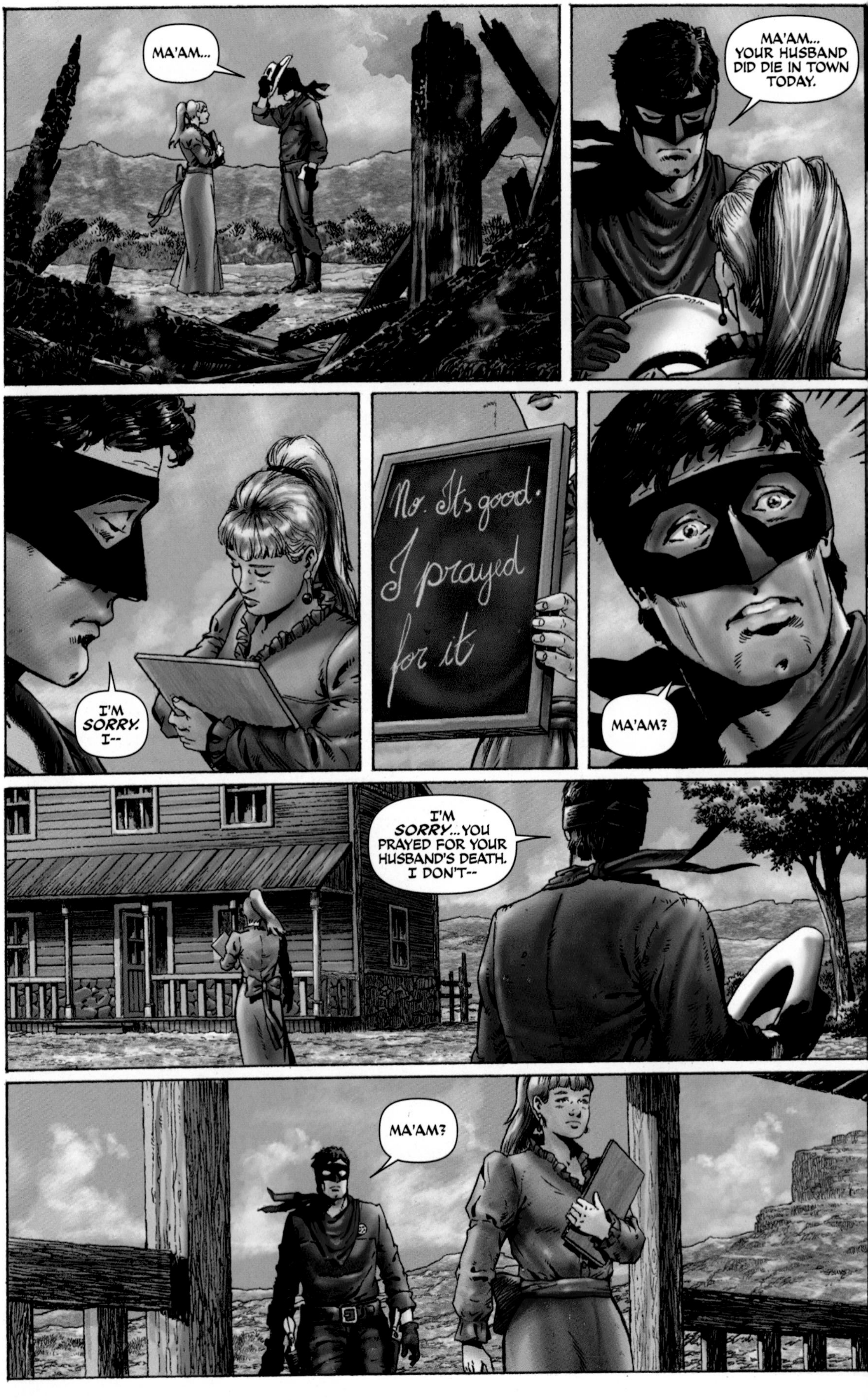
MA'AM...
MA'AM... YOUR HUSBAND DID DIE IN TOWN TODAY.
I'M SORRY. I--
No. It's good. I prayed for it
MA'AM?
I'M SORRY... YOU PRAYED FOR YOUR HUSBAND'S DEATH. I DON'T--
MA'AM?

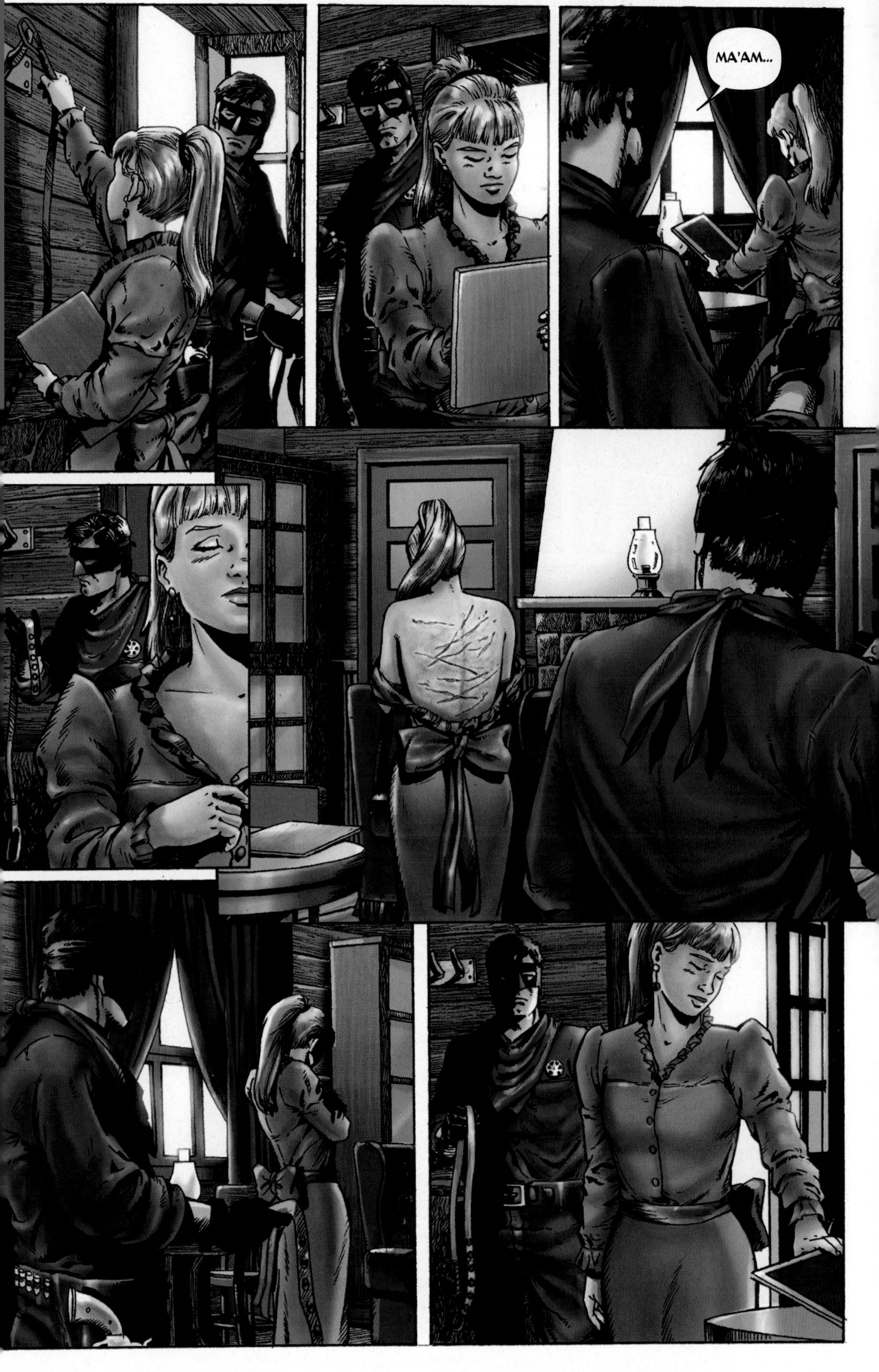
MA'AM...

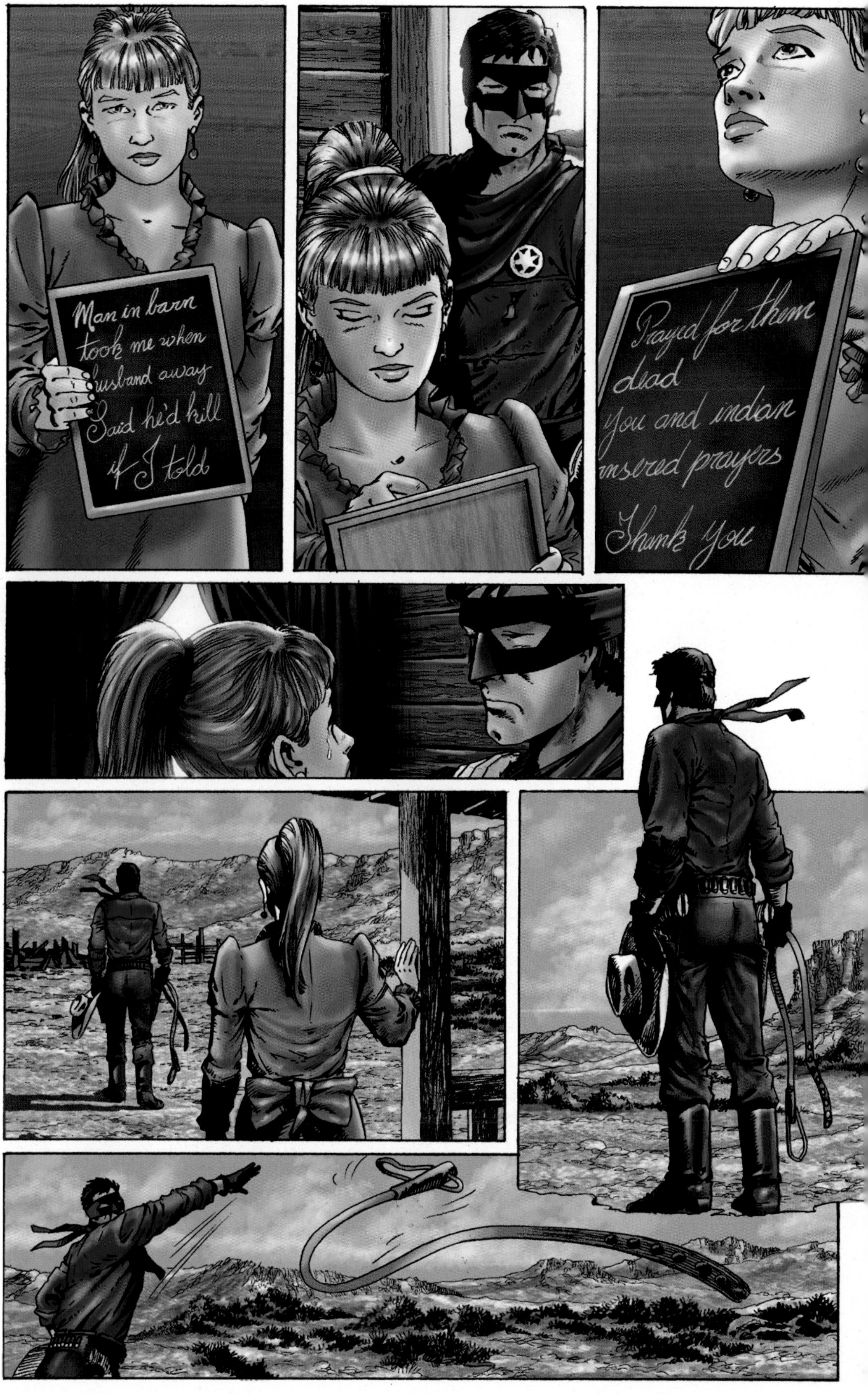
Man in barn
took me when
husband away
Said he'd kill
if I told
Prayed for them
dead
You and indian
nsered prayers
Thank you

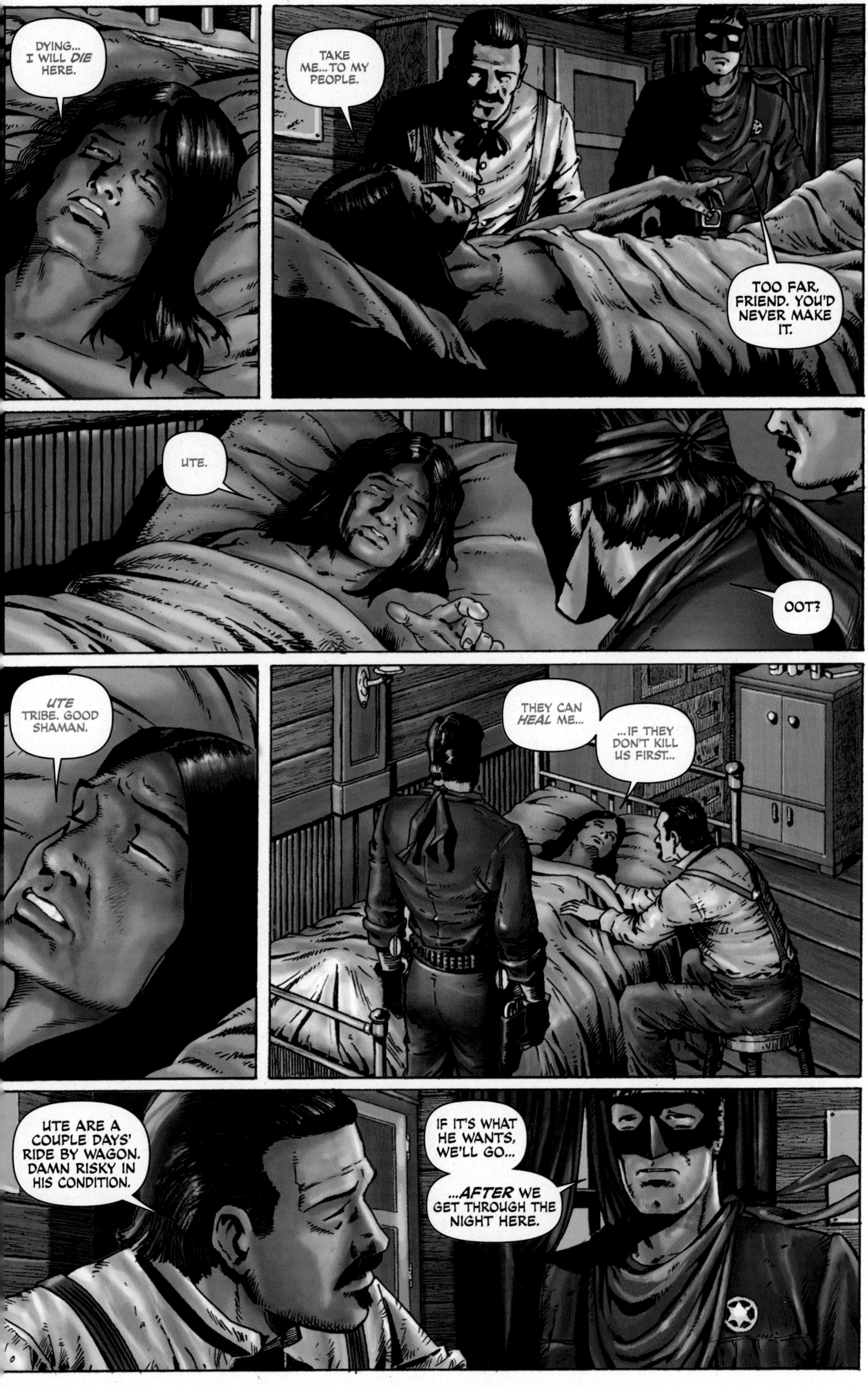
DYING... I WILL DIE HERE.
TAKE ME...TO MY PEOPLE.
TOO FAR, FRIEND. YOU'D NEVER MAKE IT.
UTE.
OOT?
UTE TRIBE. GOOD SHAMAN.
THEY CAN HEAL ME...
...IF THEY DON'T KILL US FIRST...
UTE ARE A COUPLE DAYS' RIDE BY WAGON. DAMN RISKY IN HIS CONDITION.
IF IT'S WHAT HE WANTS, WE'LL GO...
...AFTER WE GET THROUGH THE NIGHT HERE.

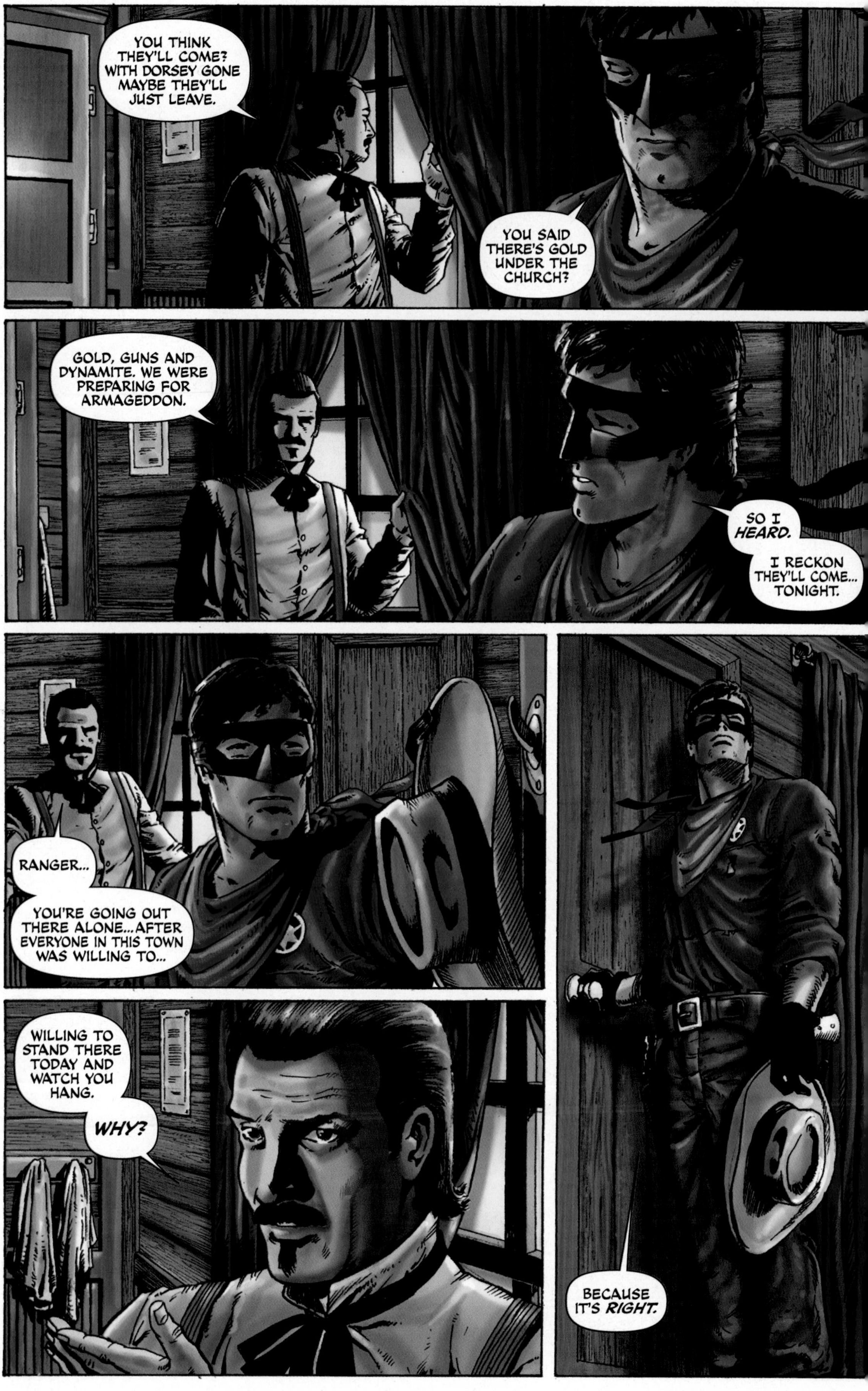
YOU THINK THEY'LL COME? WITH DORSEY GONE MAYBE THEY'LL JUST LEAVE.
YOU SAID THERE'S GOLD UNDER THE CHURCH?
GOLD, GUNS AND DYNAMITE. WE WERE PREPARING FOR ARMAGEDDON.
SO I *HEARD.*
I RECKON THEY'LL COME... TONIGHT.
RANGER...
YOU'RE GOING OUT THERE ALONE...AFTER EVERYONE IN THIS TOWN WAS WILLING TO...
WILLING TO STAND THERE TODAY AND WATCH YOU HANG.
WHY?
BECAUSE IT'S *RIGHT.*

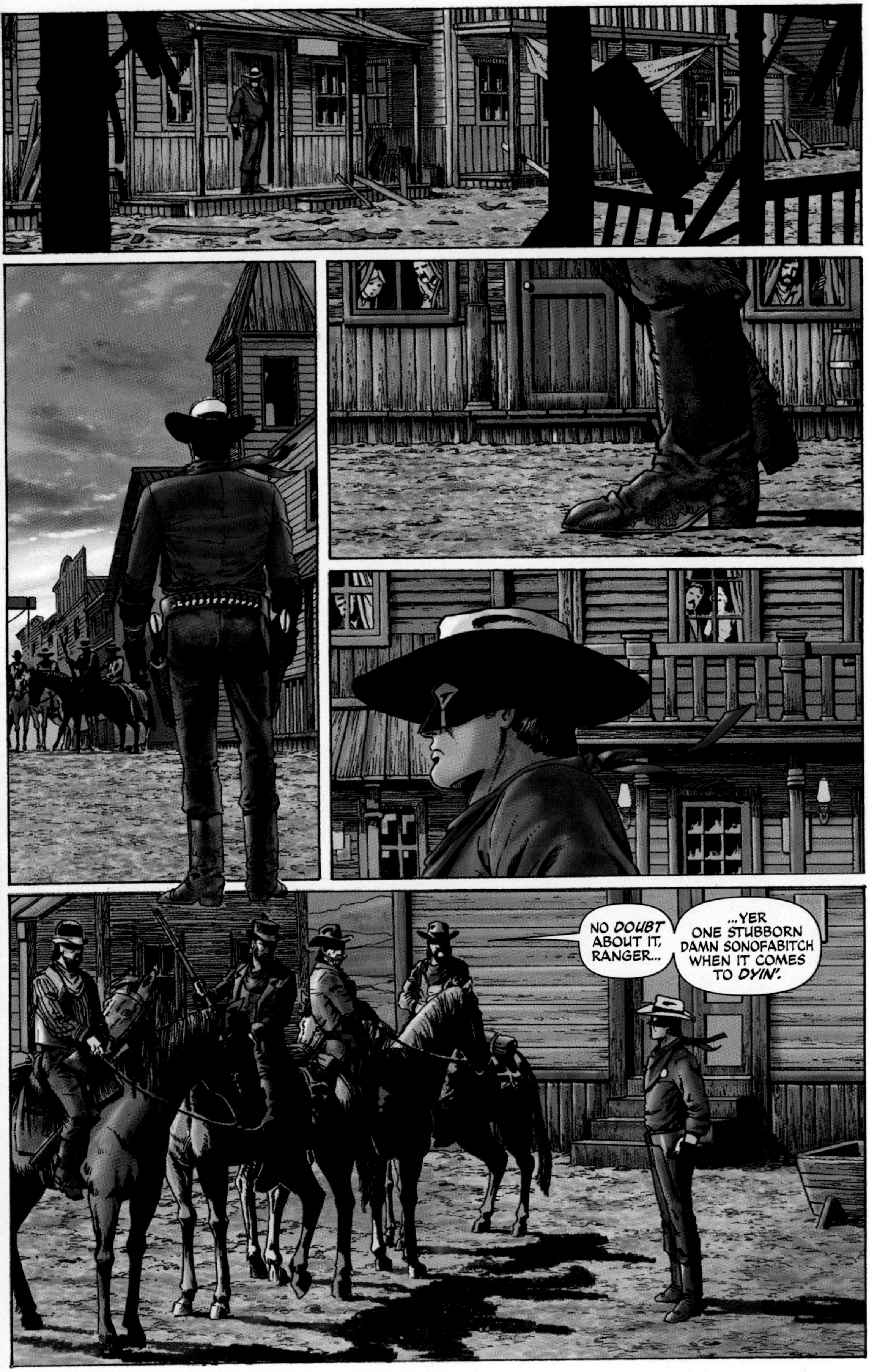
NO DOUBT ABOUT IT, RANGER...
...YER ONE STUBBORN DAMN SONOFABITCH WHEN IT COMES TO DYIN'.

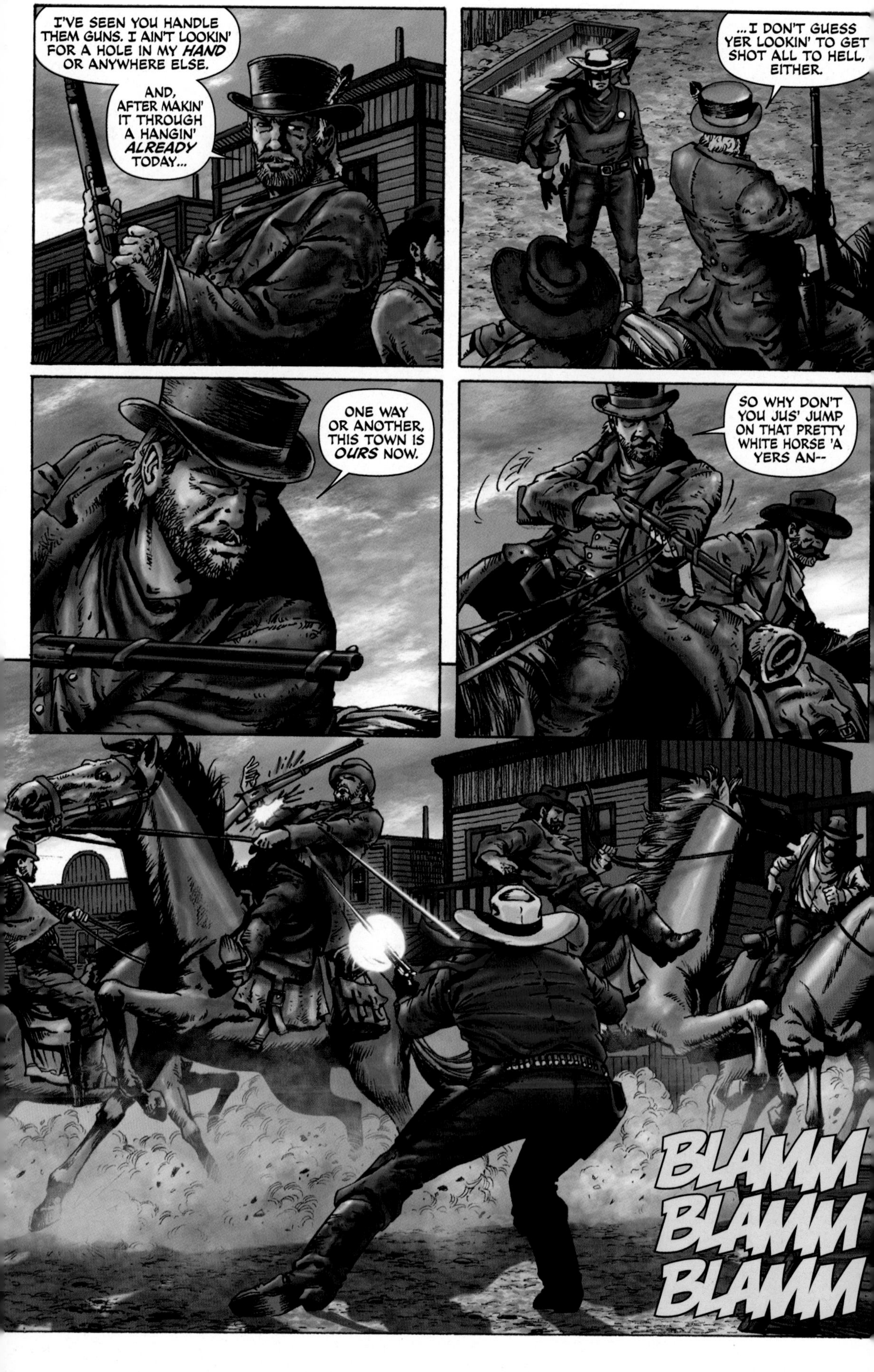
I'VE SEEN YOU HANDLE THEM GUNS. I AIN'T LOOKIN' FOR A HOLE IN MY HAND OR ANYWHERE ELSE.
AND, AFTER MAKIN' IT THROUGH A HANGIN' ALREADY TODAY...
...I DON'T GUESS YER LOOKIN' TO GET SHOT ALL TO HELL, EITHER.
ONE WAY OR ANOTHER, THIS TOWN IS OURS NOW.
SO WHY DON'T YOU JUS' JUMP ON THAT PRETTY WHITE HORSE 'A YERS AN--
BLAMM
BLAMM
BLAMM

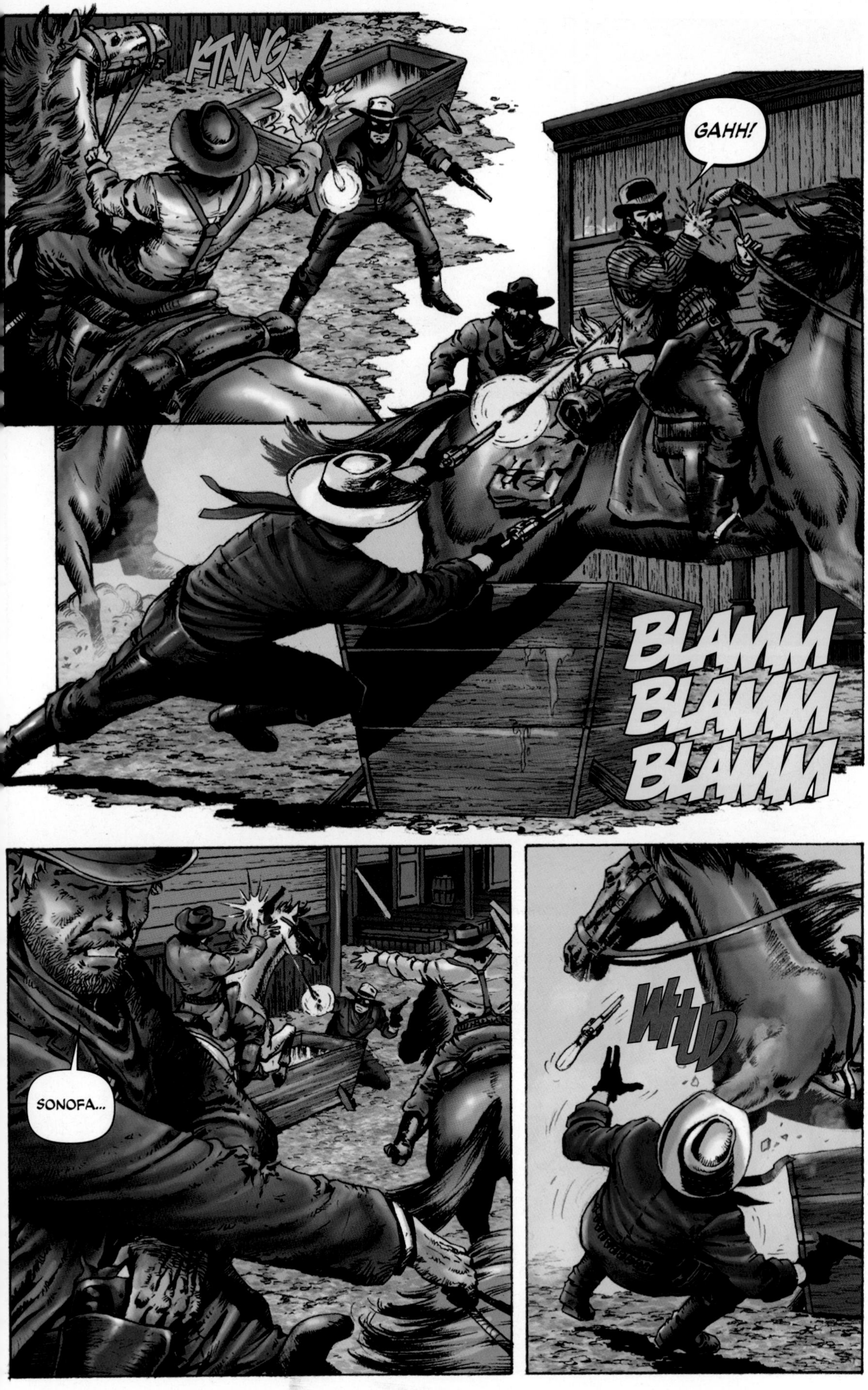
KTNNG
GAHH!
BLAMM
BLAMM
BLAMM
SONOFA...
WHUD

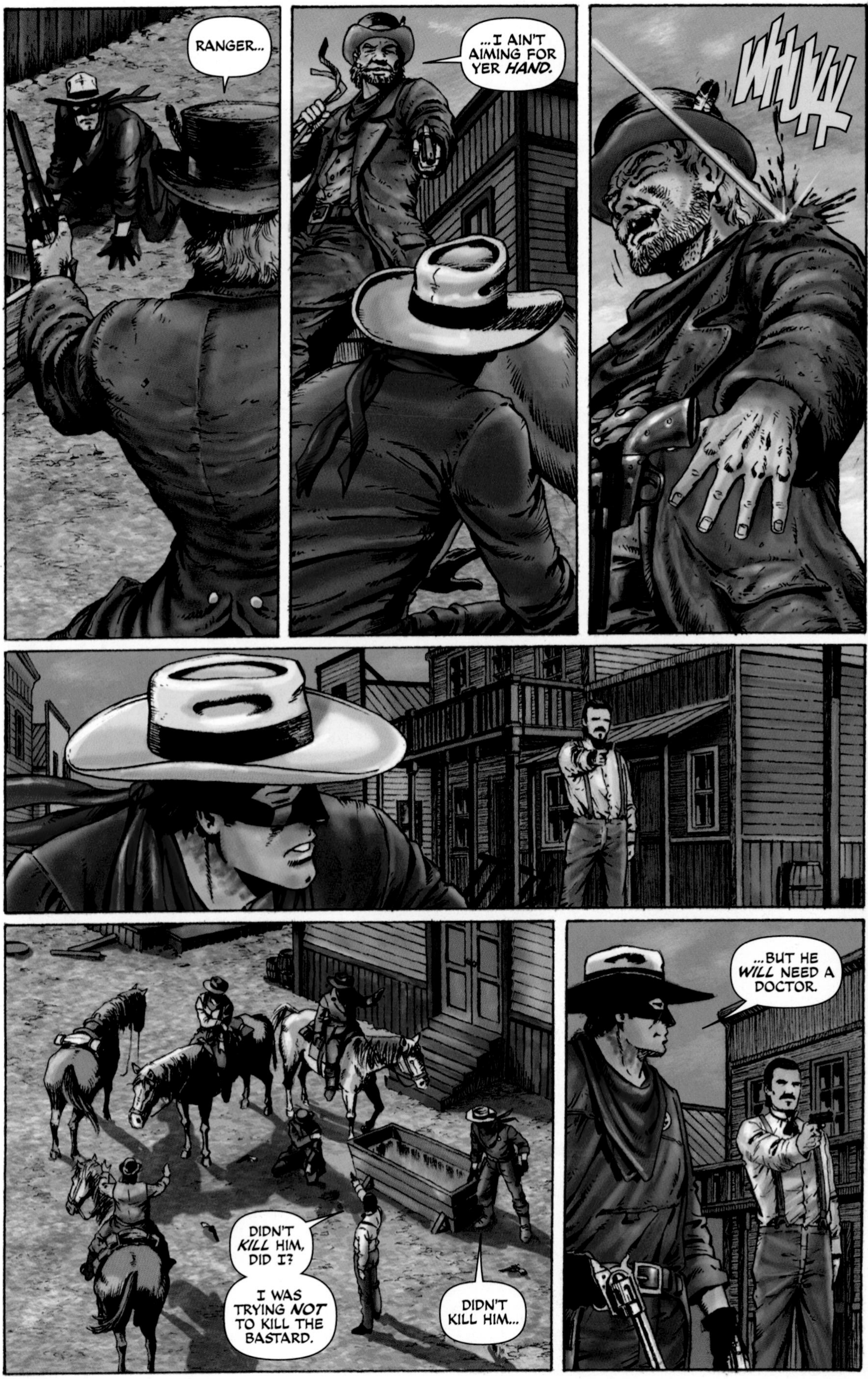
RANGER...
...I AIN'T AIMING FOR YER HAND.
WHUKK
...BUT HE WILL NEED A DOCTOR.
DIDN'T KILL HIM, DID I?
I WAS TRYING NOT TO KILL THE BASTARD.
DIDN'T KILL HIM...

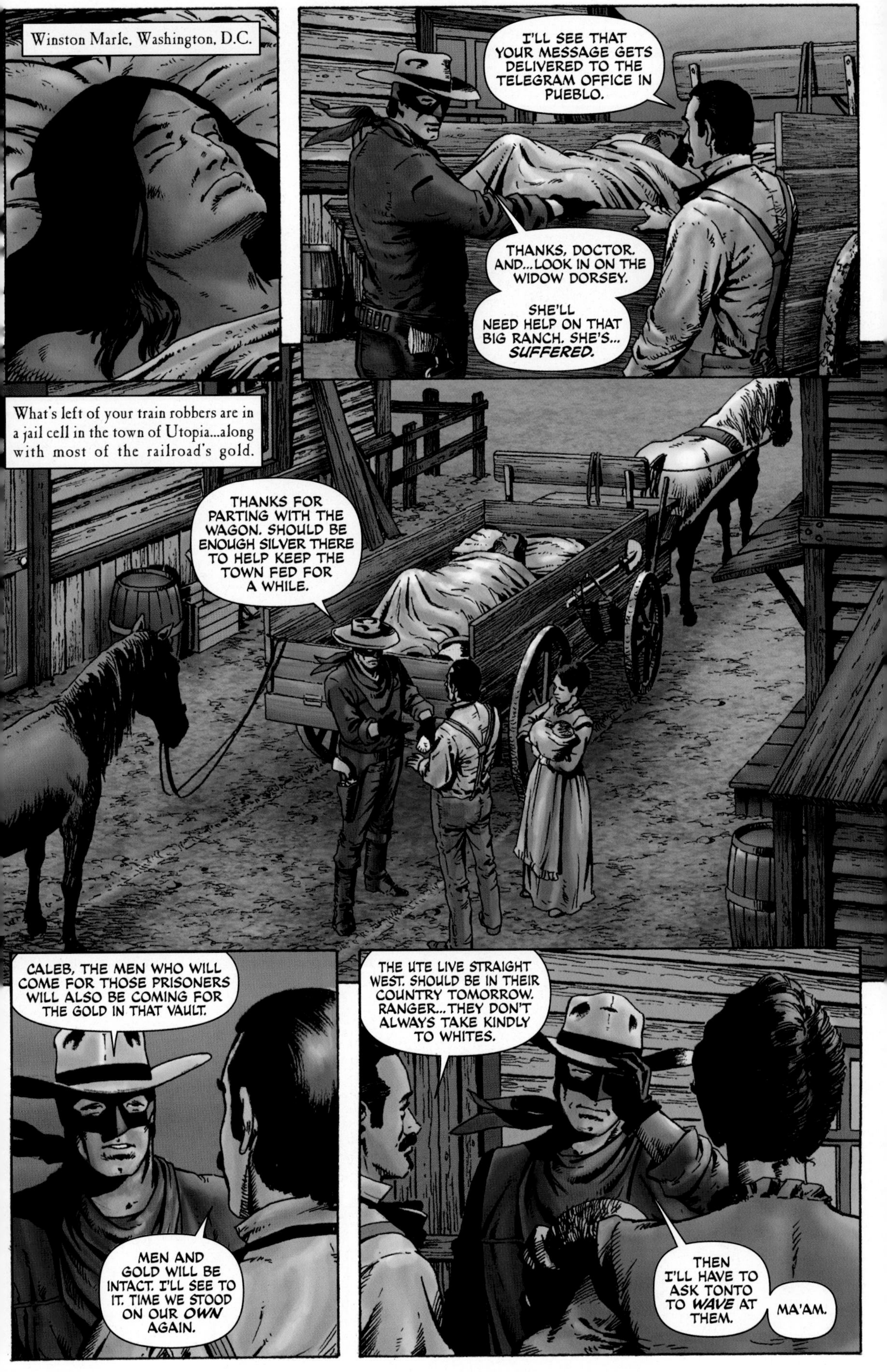

Winston Marle, Washington, D.C.
I'LL SEE THAT YOUR MESSAGE GETS DELIVERED TO THE TELEGRAM OFFICE IN PUEBLO.
THANKS, DOCTOR. AND...LOOK IN ON THE WIDOW DORSEY.
SHE'LL NEED HELP ON THAT BIG RANCH. SHE'S... SUFFERED.
What's left of your train robbers are in a jail cell in the town of Utopia...along with most of the railroad's gold.
THANKS FOR PARTING WITH THE WAGON. SHOULD BE ENOUGH SILVER THERE TO HELP KEEP THE TOWN FED FOR A WHILE.
CALEB, THE MEN WHO WILL COME FOR THOSE PRISONERS WILL ALSO BE COMING FOR THE GOLD IN THAT VAULT.
MEN AND GOLD WILL BE INTACT. I'LL SEE TO IT. TIME WE STOOD ON OUR OWN AGAIN.
THE UTE LIVE STRAIGHT WEST. SHOULD BE IN THEIR COUNTRY TOMORROW. RANGER...THEY DON'T ALWAYS TAKE KINDLY TO WHITES.
THEN I'LL HAVE TO ASK TONTO TO WAVE AT THEM.
MA'AM.

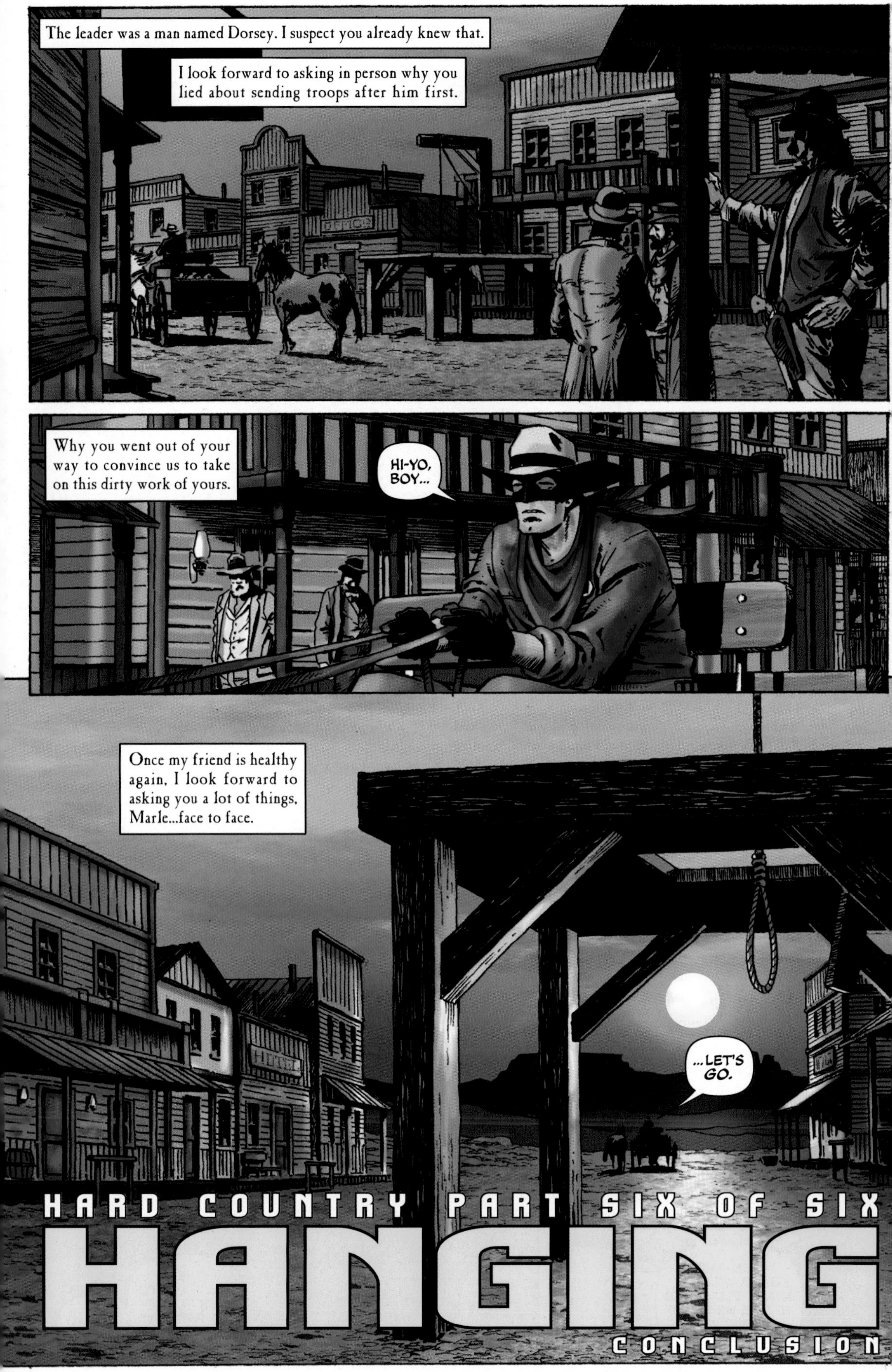

The leader was a man named Dorsey. I suspect you already knew that.
I look forward to asking in person why you lied about sending troops after him first.
Why you went out of your way to convince us to take on this dirty work of yours.
HI-YO, BOY...
Once my friend is healthy again, I look forward to asking you a lot of things, Marle...face to face.
...LET'S GO.
HARD COUNTRY PART SIX OF SIX
HANGING
CONCLUSION

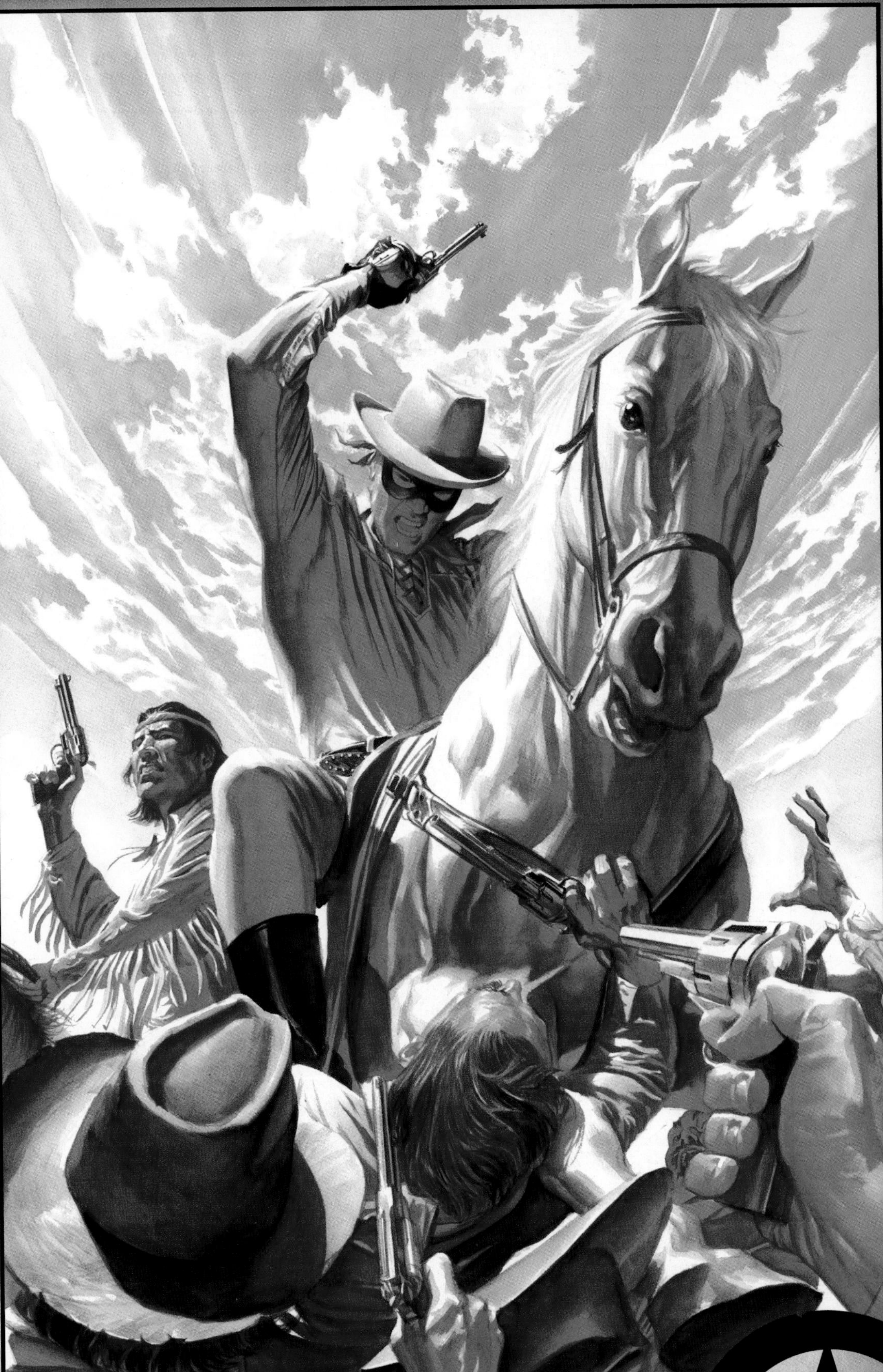

ISSUE #1 COVER BY **ALEX ROSS**

1

WRITER'S COMMENTARY

Commentary on issue #1 by **ANDE PARKS** art by **ESTEVE POLLS**

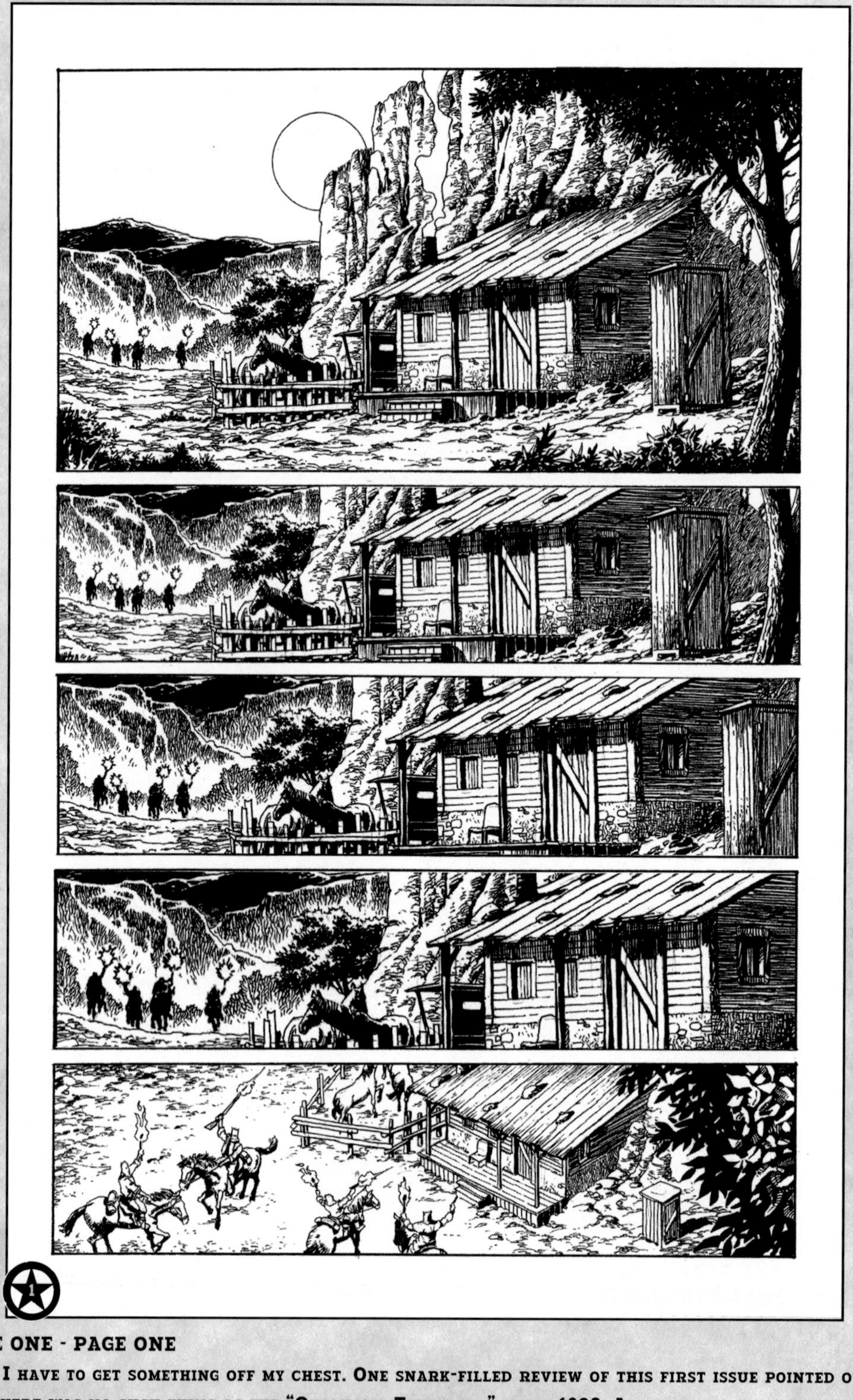

ISSUE ONE - PAGE ONE

First, I have to get something off my chest. One snark-filled review of this first issue pointed out that there was no such thing as the "Oklahoma Territory" until 1890. As a writer who prides himself on research and adding historical touches to my fiction, this hurt! I concede the point, and was aware of that inaccuracy when I wrote this first caption box. While I do place importance on research and accuracy, I also think we are here to tell a story. An engaging story about heroic fictional characters. I made the conscious decision that cheating the facts a bit was good for the story. "Oklahoma Territory, 1870" reads better and is less distracting than "The unorganized territory that would eventually become Oklahoma, 1870." That chip lifted from my shoulder... let's proceed.

It took me a long time to find a narrative device I liked for the opening of this book. I wanted to do something that got right to our narrative, but that also let the reader in on our theme: that the old west was a brutal landscape in which life was often cheap (but that a few good men could make a real difference). Once the idea of a newspaper editorial came to me, I wasted a few hours searching for an actual clipping. With the time gone forever, I made up an editorial.

As an aside, I shared my fake editorial with my pal and soon-to-be best-selling author, Alex Grecian, just to see if he thought I had captured the style of the era. Alex kindly offered a few suggestions.

Great coloring here by Marcelo Pinto. I love the torchlight against the dark blue night.

PAGE TWO
In reaching out for a name that seemed authentic for a settler from the East, I settled on the surname of my friend Dan Jurgens. Dan actually noticed and teased me about it when the book came out. I apologized for what I did to his fictional wife. We're cool now. No more name-dropping.

I wanted the death of Martha Jurgens to play very fast and chaotic... like these horrible circumstances almost couldn't be avoided. These bad guys haven't come here with murder in mind, necessarily. It just happened. That's the nature of life on the plains, far away from the nearest legal authority.

I must confess that I love that "harshly earned" line at the end of the editorial. Lincoln, KS is a real town in central Kansas. The town had a pretty well-respected newspaper by 1882.

PAGE THREE
I intentionally asked Esteve to pay homage to the final shot of "The Great Train Robber.y. He did so very well. The John Ford quote came from our fantastic editor, Joe Rybandt. I asked if he had a good quote relating to vengeance/justice. I know we're relying on a lot of narrative tricks here, between the fake editorial, the Ford quote and the upcoming letter overlay. Maybe I laid it on a bit thick. I will confess to a little "pulling out all the tricks for the first issue" syndrome.

I wanted to show Lone Ranger doing some target practice, in part to point out how hard it is for him to take down bad guys without killing them.

PAGE FOUR
I decided on this "letter as narration" device early on. It's something I've done before. I think it's a very effective tool in the modern era of comics, in which an omniscient author's voice is frowned upon stylistically. In this case it's a virtual letter... the letter Jurgens never writes down, but has composed in his head.

I DIDN'T REALIZE IT UNTIL AFTER THIS ISSUE WAS COMPLETE, BUT OUR STORY HERE BECAME A WEAPON IN MY ONGOING WAR AGAINST WHAT I THINK HAS BECOME THE EASY BUT UNREWARDING CRUTCH OF TOO MANY COMIC BOOK SCRIPTS: CLEVERNESS. I READ A LOT OF COMICS THAT LEAVE ME FEELING NOTHING BUT, "WELL, THAT WAS... CLEVER." THESE COMICS DON'T DELIVER WHAT I VALUE MOST IN ANY ART FORM: THEY DON'T MOVE ME. WITHOUT INTENDING TO MAKE A STATEMENT ON THIS TREND, I CONSTRUCTED A FIRST ISSUE FOR LONE RANGER THAT PUT THE EMPHASIS ON EMOTION OVER CLEVER. THERE IS NO TWIST ENDING HERE. THERE IS JUST, AS BEST AS I COULD DELIVER IT, A STORY OF LOSS AND HOPE, WITH OUR HERO AS THE INSTRUMENT OF CHANGE BETWEEN THE TWO.

PAGE FIVE

I ASKED A LOT OF ESTEVE ON THESE QUIET PAGES OF DESPERATION. I DIDN'T HESITATE, BECAUSE I KNOW WHAT HE'S CAPABLE OF. HE DRAWS REAL PEOPLE IN REAL ENVIRONMENTS SO WELL. HE DRAWS THEM WITH A SENSE OF STAGING A COMPOSITION THAT FEW HAVE. THESE QUIET PAGES TELL THEIR QUIET STORY WITHOUT BEING DULL. HE'S FANTASTIC. I ALSO KNOW THAT I CAN CALL FOR SETTINGS LIKE THIS CRUDE HOME WITHOUT GOING CRAZY WITH DETAIL. ESTEVE FILLS IN ENOUGH TEXTURE AND DETAIL TO MAKE THE SCENE REAL.

I AM PROUD OF JURGENS' VOICE IN THE LETTER NARRATION. IT WAS A TOUGH CALL, JUGGLING HOW I THOUGHT HE WOULD SPEAK WITH HOW HE MIGHT WRITE IF HE ACTUALLY COMPOSED THIS LETTER.

PAGE SIX

I NAMED JURGENS' DAUGHTER AFTER MY GRANDMOTHER, WHO GREW UP IN KANSAS SOME THIRTY-FIVE YEARS AFTER OUR STORY. SHE LIVED WITH DEATH IN A WAY I CAN ONLY IMAGINE. I GAVE YOUNG KATHRYN A DOLL BASED ON A DETAIL MY MOTHER PROVIDED ABOUT HER OWN CHILDHOOD. THESE CHILDREN WOULD HAVE CLUNG DESPERATELY TO THE FEW POSSESSIONS THEY COULD CALL THEIR OWN.

AMAZING COLORING OF THE FLAMES HERE BY MARCELO. THE REFLECTION OF THE FLAME AGAINST THE ROCKS IN THE FINAL PANEL IS ALSO QUITE NICE.

PAGE SEVEN

I WANTED TO CONVEY THAT THESE CHILDREN HAVE BEEN BROUGHT UP WITH SO MUCH HARDSHIP THAT THEY ALMOST DON'T HAVE TIME TO GRIEVE FOR THEIR MOTHER. THEY HAVE NOTICED THE LOSS, BUT GOING TO PIECES ABOUT IT JUST ISN'T REALISTIC. OR, THEY MIGHT JUST BE IN SHOCK. I WON'T COMPLAIN EITHER WAY YOU WANT TO SEE IT.

IT'S SUBTLE, BUT THIS LAST PANEL SHOULD RAISE THE POSSIBILITY THAT JURGENS IS CONSIDERING AN EASY WAY OUT FOR HIMSELF AND HIS CHILDREN. EARLIER VERSIONS OF THE SCRIPT HAD JURGENS MENTION THAT THEY COULD REJOIN MARTHA IN A BETTER PLACE. I THOUGHT THAT WAS TOO SPOT ON, SO I DROPPED IT.

BY THE WAY, OUR WONDERFUL LETTERER SIMON BOWLAND DID A GREAT JOB WITH THE FONT ON JURGENS' NARRATIVE.

PAGE EIGHT

I LOVE THIS PAGE. THE HINT OF SOMETHING HAPPENING IN THE FIRST PANEL, WITHOUT A SOUND EFFECT TO TIP THE ACTION. ESTEVE'S FANTASTIC REVEAL OF THE LONE RANGER. THE RANGER HOLDING SILVER'S REINS IN HIS TEETH. THE INCREDIBLE COLORING. ENOUGH. I LIKE IT.

PAGE NINE

OUR FIRST PANEL SHOWS WHAT WILL BE A RECURRING THEME IN OUR BOOK: THE LONE RANGER TRIES TO DISARM BAD GUYS BY SHOOTING THEIR GUNS OUT OF THEIR HANDS, BUT IT'S IMPOSSIBLE TO DO SO CLEANLY EVERY TIME. SOMETIMES A SERIOUS WOUND IS LEFT BEHIND. SOMETIMES THAT WOUND HAS SERIOUS CONSEQUENCES.

I LOVE THE GUY IN THE LAST PANEL ALREADY RAISING HIS HANDS IN SURRENDER. DIDN'T TAKE LONG FOR HIM TO REALIZE HE'S HAD ENOUGH. I ALSO LIKE THE OVERALL PACING HERE. THE LONE RANGER HAS DONE THIS BEFORE. HE ACTS WITHOUT HESITATION.

I FULLY REALIZE WE DIDN'T ADDRESS HOW THE LONE RANGER MIGHT HAVE KNOWN THAT HE WOULD BE NEEDED HERE AT THIS TIME. I CONSIDERED ADDING SOME STUFF ABOUT HIM GETTING WORD OF A SHOOTING THE NIGHT BEFORE, AND THEN CHECKING OUT THE FARM, HENCE BEING IN THE RIGHT PLACE AT THE RIGHT TIME. IN THE END, THOUGH, I DECIDED TO LEAVE IT MYSTERIOUS. SINCE WE'RE SEEING THIS THROUGH JURGENS' EYES, I THINK THE MYSTERY WORKS.

PAGE TEN

I KNOW TONTO GETS A SHORT SHRIFT HERE. I WANTED TO FOCUS ON THE SINGULAR VISION OF THE LONE RANGER, APPEARING TO THE JURGENS FAMILY AS A SAVIOR ON A WHITE HORSE. I THOUGHT HAVING TONTO ON THE PERIMETER OF THE ACTION, WATCHING FOR FLEEING BAD GUYS, MADE SOME STRATEGIC SENSE. REST ASSURED, NONE OF THIS SIGNALS THAT WE INTEND TO DOWNPLAY TONTO IN THIS SERIES.

PAGE ELEVEN

THIS PAGE IS ABOUT THE RANDOM, CRUEL NATURE OF JURGENS' LOSS. HIS WIFE HAS DIED OVER AN EIGHT DOLLAR DISPUTE. THIS WAS INSPIRED BY THE CLUTTER MURDERS, WHICH TRUMAN CAPOTE DOCUMENTED IN HIS MASTERPIECE, "IN COLD BLOOD." HAVING WRITTEN A GRAPHIC NOVEL ABOUT TRUMAN'S STRUGGLE TO WRITE THAT BOOK, I'M VERY FAMILIAR WITH THE SUBJECT. THE CLUTTER'S MURDERERS LEFT BEHIND FOUR BODIES. THE ROBBERY THAT STARTED THE CRIME NETTED THE KILLERS ABOUT TWENTY DOLLARS.

PAGE TWELVE

THE LONE RANGER WOULD NEVER HAVE ALLOWED JURGENS TO PULL THAT TRIGGER. STILL, HE WANTS JURGENS NOT PULLING IT TO FEEL LIKE A CHOICE.

YOU COULD SAY THIS IS THE SECOND SUCKER PUNCH FOR TONTO IN THIS ISSUE. I THINK HE'S JUST BEING HUMANE.

Page Thirteen

I doubt Tonto is thrilled about tending to these men's wounds. Such is a partnership!

More gorgeous coloring. I need to write more scenes with flames!

Here's the crux of the whole issue. The Lone Ranger has already made a difference by saving the lives of Jurgens and his family. That's meaningless, though, if Jurgens isn't strong enough to go on from here.

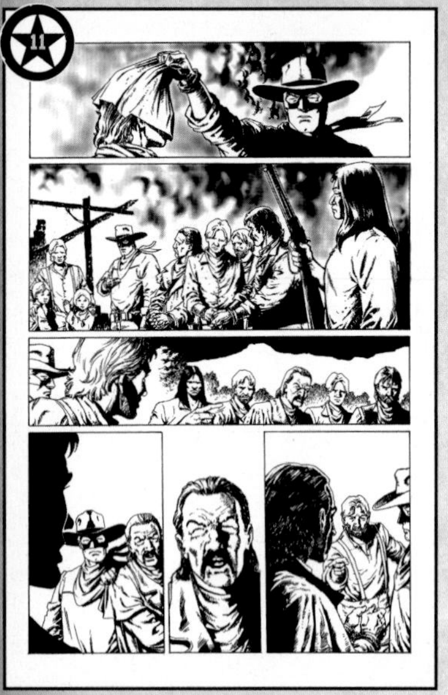

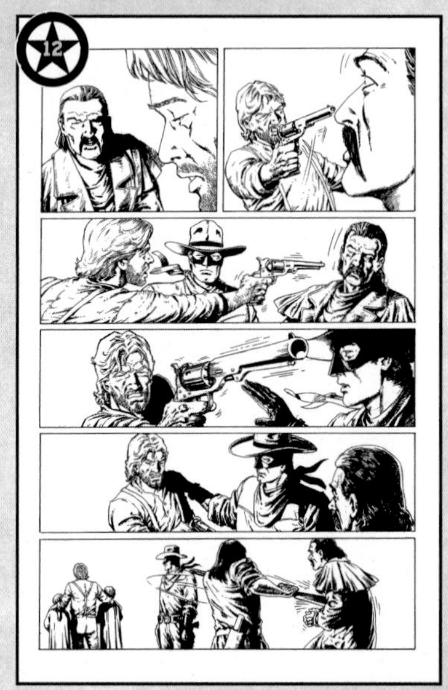

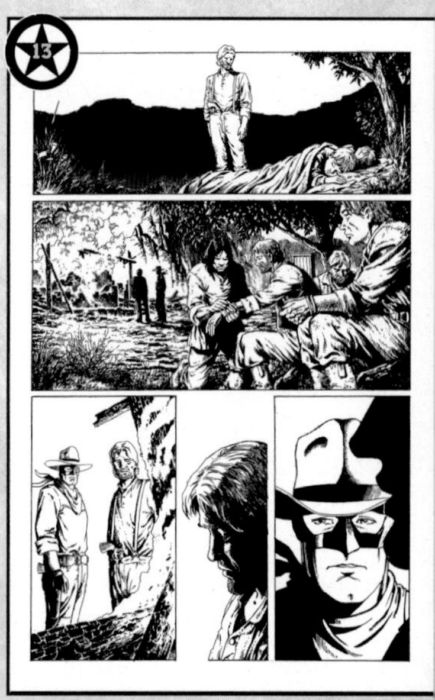

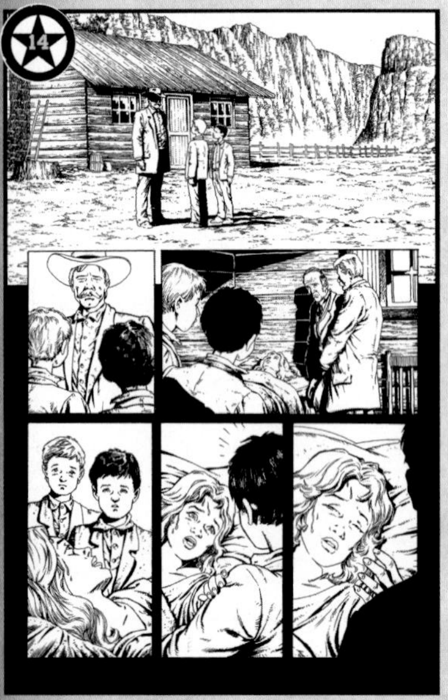

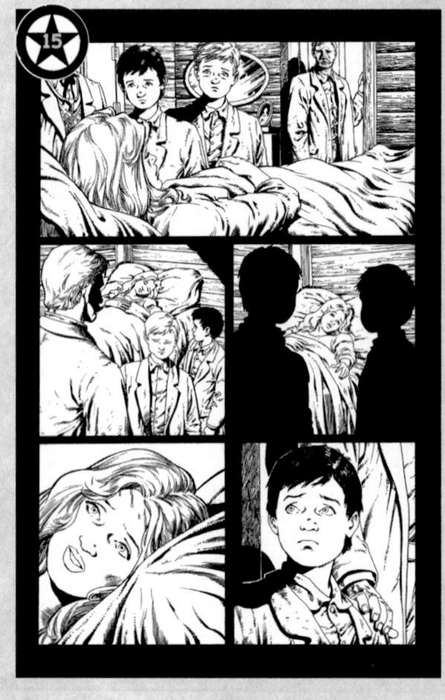

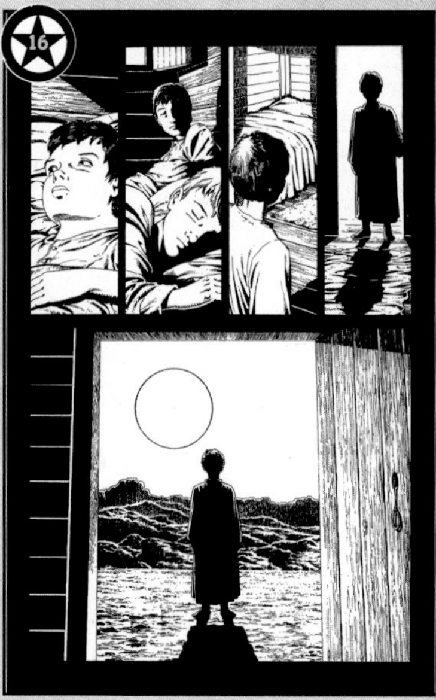

Page Fourteen

In my original pitch, each issue of our series featured a flashback sequence. We will eventually drop this premise, but I think this one works very well. We get to revisit the Lone Ranger's origins without a whole issue's recap.

This issue was built around the premise of a tragedy that mirrors the loss the Lone Ranger has faced... the loss that shaped him. Now we get to dive into the parallels.

I don't know if the dialogue of the Lone Ranger's dying mother works entirely. I wanted her to be delusional in her fever...

PAGE FIFTEEN

... UNTIL HER FINAL MOMENTS ALONE WITH HER SONS, WHERE SHE SNAPS BACK TO HERSELF FOR JUST AN INSTANT. HER DYING MOMENTS ARE OF CONCERN FOR THE MEN SHE LOVES.

I'M ALSO VERY FOND OF JAMES REID'S STOIC NATURE HERE. AS RE-IMAGINED IN JUST A FEW ISSUES BY BRETT MATTHEWS AND COMPANY, HE'S A FANTASTIC CHARACTER.

PAGE SIXTEEN

I CONFESS THAT THIS LEISURELY PACING FLIES AGAINST WHAT I GENERALLY WANT TO DO. I THINK IT'S WORTH IT FOR THAT SPLASH OF TINY JOHN REID AGAINST THAT HUGE FULL MOON. ANOTHER INTENDED HOMAGE HERE: THIS TIME IT'S THE DOORWAY AT THE END OF JOHN FORD'S "THE SEARCHERS." YOU HAVEN'T SEEN IT? GO... PLEASE. THIS COLUMN WILL STILL BE HERE WHEN YOU GET BACK.

PAGE SEVENTEEN

MAYBE JAMES REID WIPING A TEAR AT HIS WIFE'S GRAVE IS A BIT MUCH? GIVEN THE CHANCE, I MIGHT TWEAK THAT FOURTH PANEL.

I LOVE THE WAY ESTEVE FRAMED THE TWO REIDS AGAINST THE FULL MOON IN THE MIDDLE OF THE PAGE. THE COMPOSITION OF THE FINAL PANEL IS ALSO VERY NICE.

PAGE EIGHTEEN

I WANTED JAMES REID'S LAST WORD, "ALWAYS..." TO HOVER OVER THE IMAGE OF HIM BEING SHOT IN THIS FIRST PANEL. THE LONE RANGER'S ORIGIN STORY IN FOUR PANELS. THAT'S A STRONG ORIGIN. THANKS, FRAN STRIKER.

THERE'S AN OLD WRITER'S CLICHÉ THAT SAYS YOU SHOULD DUMP LINES YOU'RE ESPECIALLY PROUD OF. WELL, I'LL ADMIT THAT I'M QUITE PROUD OF THE LONE RANGER'S SPEECH ABOUT WHAT WE OWE THE DEAD THAT STARTS ON THIS PAGE. GLAD I KEPT IT.

PAGE NINETEEN

I JUST WANTED TO REFLECT ON THE LONE RANGER'S ALLEGIANCE TO HIS FATHER'S LEGACY HERE. PLUS, I KNEW ESTEVE WOULD KILL ON THE SPLASHY IMAGE AT THE GRAVE SITE. HE DID. ADD MORE KILLER COLORING AND IT ENDS UP BEING A PRETTY STRIKING PAGE.

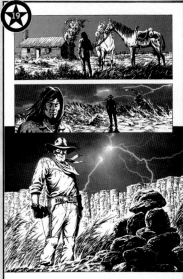

PAGE TWENTY

THIS IS WHAT MAKES THE LONE RANGER A HERO. IT'S NOT HOW WELL HE RIDES OR SHOOTS OR PUNCHES. IT'S ABOUT HIS EXAMPLE. HE IS AN UPRIGHT MAN WHO STRIVES NOT TO COMPROMISE HIS PRINCIPLES IN A WORLD THAT ASSAULTS THEM DAILY. THAT MAN CAN MAKE A DIFFERENCE. IT'S TRUE NOW AND IT WAS TRUE IN 1870.

PAGE TWENTY-ONE

With the bad guys apprehended, I had to decide what the heck the Lone Ranger and Tonto would do with them. There's no local jail to throw them into. So, a journey to real justice, which is a hard ride away. This leads directly into our next issue.

We make it clear here that the Lone Ranger is moving on, but that he's not abandoning the Jurgens family. How thorough will his follow-up be? Give us enough time and we may well get to that.

I confess to grinning like an idiot as I typed "Hi-yo, Silver!" Hell, I'm grinning writing about writing it.

PAGE TWENTY-TWO

I wasn't sure about doing this final page as a splash. There's no action... just a family looking off toward an uncertain future. Still, I wanted this moment to stand out. It is what we're here to say: the Lone Ranger has given this family hope. I think it has the impact I was hoping for. Visually, I needn't have worried. Esteve and Marcelo knocked this out of the park, and Simon's title lettering is perfect.

Is ending Nathan's narration with the arc's title too on the nose? I hemmed and hawed about it, but I'm glad I left it in there.

These stand-alone issues are tougher to write, just because you have to think of that many more ideas. I felt strongly that our first issue should be one-and-done. Our second issue will continue that trend. Starting with number three, we get into broader arcs. I hope to continue this kind of balance as we move forward.

ISSUE #2 DELETED SCENE
BY ANDE PARKS

I INITIALLY WROTE THIS FINAL PAGE FOR LONE RANGER #2. I LIKE IT, BUT I DECIDED IT WAS TOO SIMILAR TO THE FRAMING SEQUENCE FOR THE FINAL ISSUE OF THE LONE RANGER: THE DEATH OF ZORRO SERIES. I SCRAPPED IT IN FAVOR OF THE PAGE AT THE TRAIN STATION THAT MADE THE FINAL CUT. I LIKE THE NEW VERSION. IT FITS THE WOODSON CHARACTER BETTER. STILL, THIS ONE HAS ITS CHARM.

1 PAGE
THE FUTURE. A STATE FAIR IN KANSAS OR NEARBY. A YOUNGER WIFE BRINGS OLD GUN A LEMONADE. SETS IT ON A TABLE NEXT TO HIM. HE'S HERE TO MEET PEOPLE AND SELL/SIGN HIS LITTLE AUTOBIOGRAPHICAL BOOK. ONE ARM IS GONE ABOVE THE ELBOW. THE OTHER HE CAN HANDLE PRETTY WELL. A KID COMES IN, AWED. ASKING ABOUT KILLING. ASKING IF IT'S TRUE THAT GUN MET LR. THAT LR WAS REAL. GUN SMILES... FINAL LINE. WIFE IN THE SHOT. IT HAS BEEN A GOOD LIFE FOR GUN.

22.1
ESTABLISHING SHOT OF A FAIRGROUNDS IN WESTERN KANSAS, CIRCA 1900. IT'S A GOOD-SIZED FAIR FOR THE DAY: LOTS OF ANIMALS ON DISPLAY, ALONG WITH GAMES OF SKILL AND ATTRACTIONS. THERE ARE MANY ATTRACTIONS, FROM FREAKS TO SEMI-FAMOUS CHARACTERS OF THE OLD WEST. THE FAIR IS PRETTY CROWDED ON A HOT SUMMER DAY.

CAPTION
      KANSAS STATE FAIR. 1898.

22.2
IN A TENT THAT HOUSES A NUMBER OF THE ATTRACTIONS, WE SEE AN ATTRACTIVE AND WELL-DRESSED WOMAN OF ABOUT SIXTY-FIVE CARRYING A GLASS OF LEMONADE. SHE MOVES TOWARD AN OLD MAN WHO SITS IN A FOLDING CHAIR. THE MAN IS WOODSON, AGED SEVENTY-EIGHT. WE CAN'T SEE HIM WELL. WE JUST SEE AN OLD MAN IN A BLACK SUIT IN A CHAIR. THERE IS A SMALL TABLE NEXT TO WOODSON'S CHAIR. THE TABLE HOLDS A STACK OF WOODSON'S AUTOBIOGRAPHY: A SHORT, HARDBOUND BOOK OF A HUNDRED PAGES OR SO. AN ATTENDANT MANS THE TENT'S FLAP/DOOR.

WOMAN
      CLAY...

22.3
THE WOMAN HANDS WOODSON THE GLASS OF LEMONADE. WOODSON TAKES IT WITH AN OLD, SHAKING HAND.

WOMAN
      CLAY... I BROUGHT YOU A LEMONADE. IT'S TIME TO OPEN THE TENT. READY TO SELL SOME PAMPHLETS?

WOODSON
      THANK YOU, ELLEN.

WOODSON
      READY AS AN OLD MAN CAN BE.

22.4
WITH THE TENT OPEN, A LINE OF PEOPLE STREAM INTO THE TENT. FIRST UP IS AN ENTHUSIASTIC BOY OF ABOUT NINE. THE BOY IS WITH HIS MOTHER. BOTH ARE PRETTY WELL-DRESSED. THE FAIR IS A BIG DEAL. WOODSON'S WIFE IS IN THE SHOT. SHE STEPS UP A BIT HERE.

MOTHER
      THIS IS HIM, SON. THIS MAN WAS A REAL GUNFIGHTER.

WOODSON'S WIFE

> MARSHAL. MY HUSBAND WAS A MARSHAL.

WOODSON

> HELL, ELLEN... IT'S ALL RIGHT.

22.5

TIGHT ON WOODSON AND THE BOY. WE STILL HAVE NOT REALLY SEEN ONE OF WOODSON'S ARMS... KEEP IT HIDDEN FOR NOW.

WOODSON

> I WAS A MARSHAL IN THE OLD WEST, SON. A GUNFIGHTER, TOO, I GUESS.

WOODSON

> WANNA ASK ME ANYTHING? DON'T IMAGINE YOU'LL GET ANOTHER CHANCE.

BOY

> YESSIR. DID YOU EVER MEET ANYONE FAMOUS IN THE WILD WEST?

22.6

TIGHT ON WOODSON AS WE SEE, FOR THE FIRST TIME, THAT HE IS MISSING AN ARM. THE ARM LONE RANGER SHOT IN THE UPPER ARM DIDN'T SURVIVE THE INJURY. WOODSON'S JACKET IS PINNED UP TO THE SHOULDER ON THAT SIDE. WOODSON SMILES. WE CAN SEE AT LEAST PART OF HIS LOVELY WIFE IN THE SHOT, AS WELL.

WOODSON

> FAMOUS? YEAH... MET A MAN ONCE... SOME FOLKS CALLED HIM THE LONE RANGER. GOT PRETTY FAMOUS IN HIS TIME.

WOODSON

> IN FACT, I OWE THE MAN. I OWE HIM... EVERYTHING.

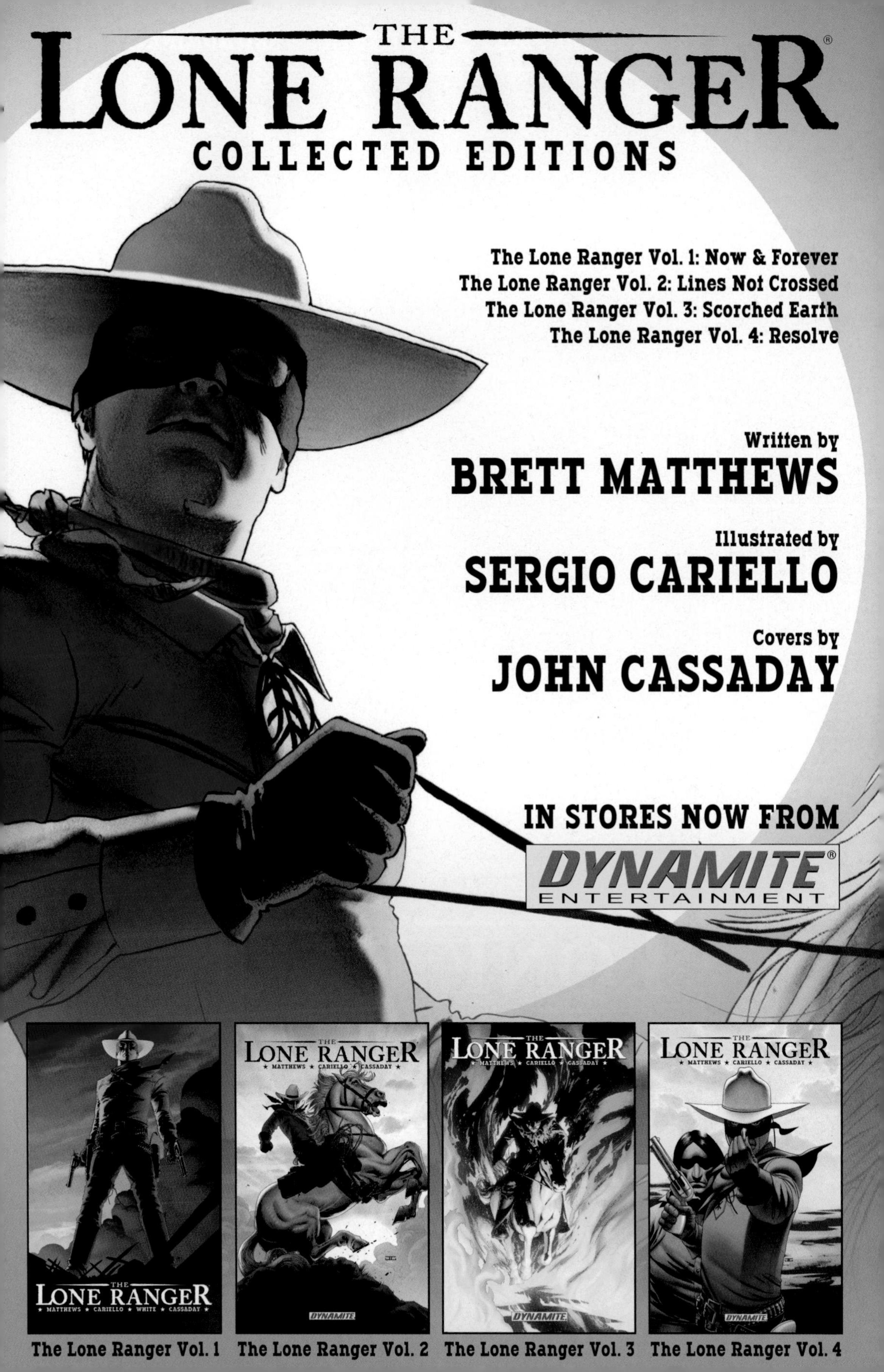
THE LONE RANGER®
COLLECTED EDITIONS
The Lone Ranger Vol. 1: Now & Forever
The Lone Ranger Vol. 2: Lines Not Crossed
The Lone Ranger Vol. 3: Scorched Earth
The Lone Ranger Vol. 4: Resolve
Written by
BRETT MATTHEWS
Illustrated by
SERGIO CARIELLO
Covers by
JOHN CASSADAY
IN STORES NOW FROM
DYNAMITE® ENTERTAINMENT
THE LONE RANGER
MATTHEWS ★ CARIELLO ★ WHITE ★ CASSADAY
THE LONE RANGER
MATTHEWS ★ CARIELLO ★ CASSADAY
DYNAMITE
THE LONE RANGER
MATTHEWS ★ CARIELLO ★ CASSADAY
DYNAMITE
THE LONE RANGER
MATTHEWS ★ CARIELLO ★ CASSADAY
DYNAMITE
The Lone Ranger Vol. 1
The Lone Ranger Vol. 2
The Lone Ranger Vol. 3
The Lone Ranger Vol. 4

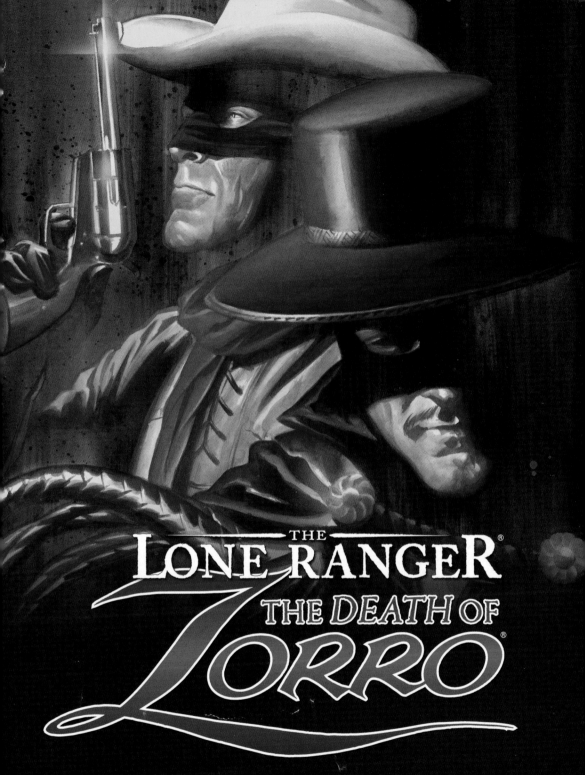

THE LONE RANGER®
THE DEATH OF ZORRO®

The legendary masked crime-fighter, Zorro, now in his sixties, has hung up his cape and sword. Living on a peaceful ranch in the new state of California, Don Diego tends to his cattle, breaks horses, and enjoys the company of his wife. When a renegade band of Confederate bushwhackers attack a nearby Indian settlement, though, Don Diego cannot just stand by and let innocents be slaughtered. Zorro heads back into action again... for the last time! Guest starring another legendary masked lawman - The Lone Ranger!

Written by **ANDE PARKS**   Art by **ESTEVE POLLS**   Cover by **ALEX ROSS**

Trade paperback in stores now!

CURRENTLY AVAILABLE AND UPCOMING COLLECTIONS FROM DYNAMITE
For a complete list, visit us online at www.dynamite.net

ARMY OF DARKNESS:

Army of Darkness: Movie Adaptation
Raimi, Raimi, Bolton

Army of Darkness: Ashes to Ashes
Hartnell, Bradshaw

Army of Darkness: Shop 'Till You Drop Dead
Kuhoric, Bradshaw, Greene

Army of Darkness vs. Re-Animator
Kuhoric, Bradshaw, Greene

Army of Darkness: Old School & More
Kuhoric, Sharpe

Army of Darkness: Ash vs. The Classic Monsters
Kuhoric, Sharpe, Blanco

Army of Darkness: From The Ashes
Kuhoric, Blanco

Army of Darkness: The Long Road Home
Kuhoric, Raicht, Blanco

Army of Darkness: Home Sweet Hell
Kuhoric, Raicht, Perez

Army of Darkness: Hellbillies & Deadnecks
Kuhoric, Raicht, Cohn

Army of Darkness: League of Light Assemble!
Raicht, Cohn

Army of Darkness Omnibus Vol. 1
Hartnell, Kuhoric, Kirkman, more

Army of Darkness Omnibus Vol. 2
Kuhoric, Raicht, Perez, more

Army of Darkness: Ash Saves Obama
Serrano, Padilla

Army of Darkness vs. Xena Vol. 1: Why Not?
Layman, Jerwa, Montenegro

Xena vs. Army of Darkness Vol. 2: What...Again?!
Jerwa, Serrano, Montenegro

Darkman vs. Army of Darkness
Busiek, Stern, Fry

BATTLESTAR GALACTICA

New Battlestar Galactica Vol. 1
Pak, Raynor

New Battlestar Galactica Vol. 2
Pak, Raynor

New Battlestar Galactica Vol. 3
Pak, Raynor, Lau

New Battlestar Galactica Complete Omnibus V1
Pak, Raynor, Jerwa, Lau

New Battlestar Galactica: Zarek
Jerwa, Batista

New Battlestar Galactica: Season Zero Vol. 1
Jerwa, Herbert

New Battlestar Galactica: Season Zero Vol. 2
Jerwa, Herbert

New Battlestar Galactica Origins: Baltar
Fahey, Lau

New Battlestar Galactica Origins: Adama
Napton, Lau

New Battlestar Galactica Origins: Starbuck & Helo
Fahey, Lau

New Battlestar Galactica: Ghosts
Jerwa, Lau

New Battlestar Galactica: Cylon War
Ortega, Nylund, Raynor

New Battlestar Galactica: The Final Five
Fahey, Reed, Raynor

Classic Battlestar Galactica Vol. 1
Remender, Rafael

Classic Battlestar Galactica Vol. 2: Cylon Apocalypse
Grillo-Marxuach, Rafael

GALACTICA 1980
Guggenheim, Razek

THE BOYS

The Boys Vol. 1 The Name of the Game
Ennis, Robertson

The Boys Vol. 2 Get Some
Ennis, Robertson, Snejbjerg

The Boys Vol. 3 Good For The Soul
Ennis, Robertson

The Boys Vol. 4 We Gotta Go Now
Ennis, Robertson

The Boys Vol. 5 Herogasm
Ennis, McCrea

The Boys Vol. 6 The Self-Preservation Society
Ennis, Robertson, Ezquerra

The Boys Vol. 7 The Innocents
Ennis, Robertson, Braun

The Boys Vol. 8 Highland Laddie
Ennis, McCrea

The Boys Vol. 9 The Big Ride
Ennis, Braun

The Boys Vol. 10: Butcher, Baker, Candlestickmaker
Ennis, Robertson

The Boys Definitive Edition Vol. 1
Ennis, Robertson

The Boys Definitive Edition Vol. 2
Ennis, Robertson

The Boys Definitive Edition Vol. 3
Ennis, Robertson, McCrea, more

THE GREEN HORNET

Kevin Smith's Green Hornet Vol. 1 Sins of the Father
Smith, Hester, Lau

Kevin Smith's Green Hornet Vol. 2 Wearing 'o the Green
Smith, Hester, Lau

Green Hornet Vol. 3 Idols
Hester, Lau

Green Hornet Vol. 4 Red Hand
Hester, Smith, Vitorino, more

Kevin Smith's Kato Vol. 1 Not My Father's Daughter
Parks, Garza, Bernard

Kevin Smith's Kato Vol. 2 Living in America
Parks, Bernard

Green Hornet: Blood Ties
Parks, Desjardins

The Green Hornet: Year One Vol. 1 The Sting of Justice
Wagner, Campbell, Francavilla

The Green Hornet: Year One Vol. 2 The Biggest of All Game
Wagner, Campbell

Kato Origins Vol. 1 Way of the Ninja
Nitz, Worley

Kato Origins Vol. 2 The Hellfire Club
Nitz, Worley

The Green Hornet: Parallel Lives
Nitz, Raynor

The Green Hornet Golden Age Re-Mastered
Various

PROJECT SUPERPOWERS

Project Superpowers Chapter 1
Ross, Krueger, Paul, Sadowski

Project Superpowers Chapter 2 Vol. 1
Ross, Krueger, Salazar

Project Superpowers Chapter 2 Vol. 2
Ross, Krueger, Salazar

Project Superpowers: Meet The Bad Guys
Ross, Casey, Lilly, Lau, Paul, Herbert

Black Terror Vol. 1
Ross, Krueger, Lilly

Black Terror Vol. 2
Ross, Hester, Lau

Black Terror Vol. 3 Inhuman Remains
Ross, Hester, Reis, Herbert

Death-Defying 'Devil Vol 1
Ross, Casey, Salazar

Masquerade Vol. 2
Ross, Hester, Laul

RED SONJA

Adventures of Red Sonja Vol. 1
Thomas, Thorne, More

Adventures of Red Sonja Vol. 2
Thomas, Thorne, More

Adventures of Red Sonja Vol. 3
Thomas, Thorne, More

Queen Sonja Vol. 1
Ortega, Rubi

Queen Sonja Vol. 2 The Red Queen
Nelson, Herbert

Red Sonja She-Devil With a Sword Vol. 1
Oeming, Carey, Rubi

Red Sonja She-Devil With a Sword Vol. 2: Arrowsmiths
Oeming, Rubi, more

Red Sonja She-Devil With a Sword Vol. 3: The Rise of Kulan Gath
Oeming, Rubi, more

Red Sonja She-Devil With a Sword Vol. 4: Animals & More
Oeming, Homs, more

Red Sonja She-Devil With a Sword Vol. 5: World On Fire
Oeming, Reed, Homs

Red Sonja She-Devil With a Sword Vol. 6: Death
Marz, Ortega, Reed, more

Red Sonja She-Devil With a Sword Vol. 7: Born Again
Reed, Geovani

Red Sonja She-Devil With a Sword Vol. 8: Blood Dynasty
Reed, Geovani

Red Sonja She-Devil With a Sword Vol. 9: War Season
Trautmann, Geovani, Berkenkotter

Red Sonja She-Devil With a Sword Omnibus Vol. 1
Oeming, Carey, Rubi, more

Red Sonja vs. Thulsa Doom Vol. 1
David, Lieberman, Conrad

Savage Red Sonja: Queen of the Frozen Wastes
Cho, Murray, Homs

Red Sonja: Travels
Marz, Ortega, Thomas, more

Sword of Red Sonja: Doom of the Gods (Red Sonja vs. Thulsa Doom 2)
Lieberman, Antonio

Red Sonja: Wrath of the Gods
Lieberman, Geovani

Red Sonja: Revenge of the Gods
Lieberman, Sampere

Savage Tales of Red Sonja
Marz, Gage, Ortega, more

VAMPIRELLA

Vampirella Masters Series Vol 1 Grant Morrison & Mark Millar
Morrison, Millar, more

Vampirella Masters Series Vol 2 Warren Ellis
Ellis, Conner Palmiotti, more

Vampirella Masters Series Vol 3 Mark Millar
Millar, Mayhew

Vampirella Masters Series Vol 4 Visionaries
Moore, Busiek, Loeb, more

Vampirella Masters Series Vol 5 Kurt Busiek
Busiek, Sniegoski, LaChanc, more

Vampirella Masters Series Vol 6 James Robinson
Robinson, Jusko, more

Vampirella Archives Vol 1
Various

Vampirella Archives Vol 2
Various

Vampirella Archives Vol 3
Various

Vampirella Archives Vol 4
Various

Vampirella Vol. 1 Crown of Worms
Trautman, Reis, Geovani

Vampirella Vol. 2 A Murder of Crows
Trautman, Jerwa, Neves, more

Vampirella And The Scarlet Legion
Harris, Malaga

MORE FROM GARTH ENNIS

Dan Dare Omnibus
Ennis, Erskine

Garth Ennis' Battlefields Vol. 1: The Night Witches
Ennis, Braun

Garth Ennis' Battlefields Vol. 2: Dear Billy
Ennis, Snejbjerg

Garth Ennis' Battlefields Vol. 3: The Tankies
Ennis, Ezquerra

Garth Ennis' The Complete Battlefields Vol. 1
Ennis, Braun, Snejbjerg, Ezquerra

Garth Ennis' Battlefields Vol. 4: Happy Valley
Ennis, Holden

Garth Ennis' Battlefields Vol.5: The Firefly and His Majesty
Ennis, Ezquerra

Garth Ennis' Battlefields Vol.6: Motherland
Ennis, Braun

Garth Ennis' The Complete Battlefields Vol. 2
Ennis, Braun, Holden, Ezquerra

Jennifer Blood Vol. 1: A Woman's Work Is Never Done
Ennis, Batista, Baal, more

Just A Pilgrim
Ennis, Ezquerra

Seven Brothers Omnibus
Ennis, Woo, Kang, more

ART BOOKS

The Art of Howard Chaykin

The Art of Red Sonja

The Art of Vampirella

The Dynamite Art of Alex Ross

George Pérez, Storyteller

The Romita Legacy